The Friction of Life

Lorenza Foschini

The Friction of Life

An investigation on Renato Caccioppoli's life

Translated by Alastair McEwen

 Springer

Lorenza Foschini
Napoli, Italy

Translated by
Alastair McEwen
Cupar, UK

ISBN 978-3-031-65261-5   ISBN 978-3-031-65262-2   (eBook)
https://doi.org/10.1007/978-3-031-65262-2

Translation from the Italian language edition: "L'attrito della vita. Indagine su Renato Caccioppoli, matematico napoletano" by Lorenza Foschini, © La nave di Teseo 2022. Published by La nave di Teseo. All Rights Reserved.

This Springer imprint is published by the registered company Springer Nature Switzerland AG
The registered company address is: Gewerbestrasse 11, 6330 Cham, Switzerland

If disposing of this product, please recycle the paper.

"The theory of normal families of multiple complex variables is dominated by Caccioppoli's theorem: a normal family with respect to every complex variable is also normal with respect to the set of these variables." Paul Montel, *Encyclopedie française*

"We are of the tribe that asks questions, and we ask them to the bitter end. Until no tiniest chance of hope remains to be strangled by our hands. We are of the tribe that hates your filthy hope, your docile, female hope!" Jean Anouilh, Antigone

In memory of the Caccioppoli sisters: Colomba, Gemma, Rosamaria and of Isabella, my mother

9934 is an asteroid discovered in 1985 by the American astronomer Edward Bowell. It has a diameter of 7.8 kilometres and has an orbit characterized by a semimajor axis equal to 2.58 astronomical units and an eccentricity of 0.23, inclined at 16.62 degrees to the ecliptic.

It is called Caccioppoli in honour of the mathematician Renato but also of his distant cousin, Francesco, who dedicated his whole life to the observation of the sky. Director of the Naval Institute of Procida, he died of pneumonia contracted after having spent a cold night at the telescope to observe the passage of a comet. It was 1904. The same year that Renato was born.

Genealogical Traces

Giuseppe Caccioppoli

son of Domenico* and Stanislaa Borzelli

Naples 11 December 1852–5 June 1947

his first marriage was to Angelina Amendola, who died on 17 March 1885, with whom he had a son, Nicola, who died at the age of 16 on 19 July 1897. His second marriage took place on 22 July 1903 with Giulia Sofia Bakunin, and from their union came:

Renato Caccioppoli

Naples 20 January 1904–8 May 1959 married Sara Mancuso on 29 June 1939

Ugo Caccioppoli

Naples 18 January 1905–18 August 1992 married Paola Trapani on 30 April 1959

Francesco Caccioppoli

son of Lorenzo* and Fortunata Savarese
Ponza 1855–Naples 6 May 1904
married Isabella Starace, from their union:
Lorenzo, Sirius, Arcturus, Perseus, Francis
were born

Lorenzo Caccioppoli

Procida 1885–Rome 1953
married Rosa Starace, from their union came:
Isabella, Colomba, Gemma, Rosamaria
(Isabella: Savona 28 May 1918–Rome 2 June 2010
married Nicola Foschini on 23 April 1939)
*Domenico Caccioppoli, father of Giuseppe, and Lorenzo Caccioppoli, father
of Francesco, were cousins

Mikhail Alexandrovich Bakunin

Priamukhino 30 May 1814–Bern 1 July 1876
married in Tomsk on 17 October 1858 Antonia* Kwiatkowska
White Russia 1840–Portici 2 June 1887
From their union were born:

Carlo

Geneva 25 May 1868–1 May 1944 married Maria Canetto and had three
children

Giulia Sofia

Orselina 17 January 1870–Naples 19 February 1956
married Giuseppe Caccioppoli and had two children, Renato and Ugo

Maria

Krasnoyarsk 2 February 1873–Naples 17 April 1960
married Agostino Oglialoro Todaro
*After Bakunin's death, Antonia married Carlo Gambuzzi

Acknowledgements

I thank Carlo Sbordone, who was the first person I contacted many years ago, telling him about my desire to write about Caccioppoli and who read the drafts of this book very carefully.

Lucio Carbone, who opened the doors for me of the institute where Renato taught and followed me during the writing of the text, patiently and helpfully providing me with suggestions and fundamental contributions.

Luigi Marino, who with extraordinary generosity took me to Maria Del Re's house, he had me read the correspondence between her and Caccioppoli, he made many photographs of Renato available and pointed out articles and publications valuable for my research.

Ettore Perozzi, astronomer, scholar of the planets. He provided me with essential information on a subject I knew nothing about.

Carmine Colella, for the documented information on Mikhail and Maria Bakunin.

Giuseppe Girimonti Greco, for his generous linguistic advice but not only that. He dispensed important information and suggestions.

Giuseppe Marcenaro, once again an invaluable source of reflections and suggestions.

Wanda Monaco. Her contribution was essential for the knowledge of Caccioppoli.

Rossana Rummo, general director of the Archives during the period of my research.

Eugenio Lo Sardo, then director of the State Archives in Rome.

Caterina Arfé, of the State Archives in Rome.

Patrizia Rusciani, director of the Library of Modern and Contemporary History of Rome.

Marinella Galateria, who introduced me to the letter Paola Masino wrote to her mother.

Daniele Conti, who told me about Sofia Bakunin's letter to Adele Croce.

Titti Marrone, for following me from the beginning and having made it possible to have a photograph of Sara Mancuso. And for this I thank Massimiliano Marotta, Sara's grandson, who generously gave it to me.

Renato Fiorenza, for his touching testimony.

Pierre Masson, president of the Association des Amis d'André Gide.

Luciano De Crescenzo, who a few years before his death spoke to me for a long time about Caccioppoli whose student he had been, and whose memories I have reported in some pages.

Salvatore Pica, who put me in contact with many people in Naples who were fundamental to my research.

Maestro Domenico Sapio, who showed me the beauties of the conservatory of San Pietro a Majella.

Raimonda Gaetani, for her youthful memories of Renato.

In this regard, my heartfelt thanks go out to a dear friend who is no longer here. Annamaria Palermo was an inexhaustible source of amazing stories, including many about Caccioppoli. To her goes my remembrance full of nostalgia and of affection.

Special thanks to Mario Andreose, Pierluigi Battista, and Felice d'Alfonso del Sordo. And again to Annamaria Boniello, Isabella Ducrot, Sergio Cappelli, Giuseppe de Lollis, Federico Forquet, Paolo Franchi, Nicoletta Lazzari, Teresa Leo, Silvano Materasso, Francesco Morabito, Duccio Trombadori, Francesco Villari, and in particular to my sister Mariella Foschini for the family mementoes and timely observations, and to my daughter Camilla Paternò for her reflections and the sensitivity with which she read these pages. As always I am grateful to my husband Marco Molendini, who patiently follows me and encourages me.

Introduction

Family Memories

Great-grandmother Isabella Starace, mother of my grandfather Lorenzo, had many children with her husband Francesco Caccioppoli, astronomer, atheist and socialist. Except for the first, Lorenzo, they all bore the name of a star: Sirio the first, who died in swaddling clothes, then Sirio the second, Perseo, Arturo the first, who died at 1 year old, Arturo the second, Elettra and Gemma, who flew to heaven as children. A ninth was on the way when Francesco fell seriously ill due to a comet that was supposed to pass through one night in April. He had remained hours and hours on the terrace above the building in Largo Ferrandina in Naples, oblivious to the unusual cold of that spring night, motionless at the telescope until, catching it by surprise, he spotted it. He went back indoors at the crack of dawn, happy and excited, with that wonderful sight inside him. Shivering and icy cold, he slipped into bed. Sleepily, great-grandmother asked him in dialect: "Francé, where have you been?"—and he replied: "Isabé, I have been with God!"

That night proved fatal for him. Peppino Caccioppoli, a great Neapolitan doctor, was called urgently. A man who had known the pain of losing his wife and 16-year-old son, Nicola, to a disease that had proved impossible to defeat. In 1903 he remarried and on 20 January 1904 he again became the father of a son, Renato. On arriving at the astronomer's bedside, the professor diagnosed a bad, very bad case of pneumonia. Chances of survival—he declared—were few. When the dying man realized that there was no longer any hope, he turned to his cousin and asked him to swear that he would take his firstborn Lorenzo under his tutelage and have him study medicine, anticipating that a solid profession would ensure he could provide for the family. This he said in a whisper. Giuseppe Caccioppoli gave his word. Francesco, trusting, closed his eyes forever.

Great-grandmother Isabella, with her children, left the house in Naples. She moved to Vico Equense, to Palazzo Starace, where she had inherited an apartment on the second floor. After a few months, in September, Francesco was born, named after his father but nicknamed Ciccillo by everyone.

Over the years, Uncle Peppino never failed to give his support. Francesco's children grew up intelligent, lively, extravagant. One day Arturo, in a stubborn attempt to accomplish an impossible feat—to cut with a razor sharp knife a slice of cheese that absolutely had to be wafer thin—he gave himself an extremely deep cut in the hand. For years he talked about this with profound emotion, recounting how Peppino, called in to help, arrived in Vico from Naples at night in a wheelchair to sew up that little hand with his renowned stitch, known as the "Caccioppoli suture".

Once my grandfather Lorenzo passed his high school diploma, as agreed, he returned to Naples and began attending the faculty of Medicine. He was a guest of his uncle in Villino Caccioppoli in Capodimonte. The surgeon lived there with his second wife Giulia Sofia Bakunin, one of the three children of Mikhail, the famous anarchist. Sofia was a doctor, among the first lady doctors in Italy, she was cosmopolitan, emancipated, and a polyglot like her father. Lorenzo was happy and curious to meet her walking through the spacious rooms of the villa overlooking the gulf, when early in the morning he was preparing to walk down Via Vecchia San Rocco to go to the university. He followed all the lessons diligently and studied with scrupulous care to become a good doctor. But soon it was time for the first anatomy lesson. Unprepared for what awaited him, he was shocked by that spectacle: faced with the corpse, the blood, his vision became blurred, his mind too, and he fell to the floor in a faint. When he showed up the next day before the great doctor, he couldn't even explain the details of what had happened. The first words stuttered with shyness mixed with fear sufficed: "Uncle, try to understand, I do not want to, I cannot, study medicine, diseases, blood, death horrify me..."

The professor, who was a gentle and reasonable man, interrupted him with unexpected violence, thundering: "I swore on your father's deathbed that I would make a doctor of you. How dare you ask me to become a perjurer? Get out of here and never darken my doorstep again!"

After the tumultuous conversation with his uncle, Lorenzo left the villa knowing that he would never return and that he would never see its occupants ever again: the surgeon with the long flowing beard, his exotic wife and their two children, Renato and the youngest, Ugo, born the following year.

As time passed, my grandfather became a successful engineer specialized in energy issues, a brilliant man in his own way, too. The memory of Giuseppe Caccioppoli and his family faded, but a subtle regret always lingered on for those bonds severed so violently. A regret that was even passed down to his daughters and then to us grandchildren when it became known that Uncle Peppino's firstborn, Renato, had become a famous mathematician surrounded by the fascinating aura of his Russian ancestor and who the people of Naples called 'o Genio, the Genius.

I have heard this story many times in my family, and I have felt the need to tell it. It wasn't easy. People know everything and nothing about Caccioppoli. And it is difficult to reconstruct his life by separating the real facts from the many legends that sprang up around him. Renato was not a misanthrope, on the contrary. He appeared affable, sociable, even worldly. A great talker and the author of mordant, cutting witticisms. But when he was seized by acute crises that bordered on desperation, he took refuge in alcohol. No one has ever managed to truly penetrate his inner world. The unfathomable universe of his genius will forever remain a mystery.

Contents

1

Teatro San Carlo
October 1958

He is standing, wrapped in a raincoat held at the waist by a sloppily knotted belt. A newspaper is sticking out from his pocket, *l'Unità* or *Paese Sera*. He is leaning against one of the white and gold consoles in the foyer of the San Carlo, crowded with the usual mix of people from high society and refined connoisseurs. It's October, concert season.

He gets my attention. He hasn't taken his eyes off my mother who, a few steps away from him, on feeling observed, clutched my hand as if I, small as I was, could relieve her of embarrassment.

I watch. A tuft of hair falls repeatedly over his forehead and covers his right eye, but with a mechanical gesture he pushes it back with his hand reversed, palm upwards, revealing a lucid, restless gaze, full of curiosity and above all a mocking, ill-concealed irony that it is also reflected in the half smile, accentuated by the cigarette held between his lips. He is quite tall and of a ghostly thinness, but he doesn't scare me. Quite the contrary.

It's not the first time I've seen him. Mum has often pointed him out to me on Sunday morning when, on leaving our house in via dei Mille, we come across him as he heads to the Alhambra cinema, where he hosts a film club frequented by passionate film buffs. His presentations, studiedly improvised, arouse the devoted admiration of his enthusiastic admirers. Even on those occasions, on coming across her, he looks at her with the same fixity as now. He is perfectly aware of who she is, he knows her name and her father's name. They have the same surname, but what attracts him even more perhaps is her beauty. Slender, with a severe elegance, so little Neapolitan with long blonde hair worn in a chignon and a blue gaze that is vague and evasive owing to

L. Foschini, *The Friction of Life*, https://doi.org/10.1007/978-3-031-65262-2_1

shyness (Fig. 1.1). She is also very beautiful tonight, in her black velvet dress and with pearl earrings on her tiny earlobes.

It's the first time I've been taken to a concert, but it's a unique opportunity. Igor Stravinsky is conducting the orchestra.

My sister, who is 16 years old, in a fit of adolescent defiance, whispers to my mother: "But don't you see how he's looking at us? Now I'm going up to him. Where's the harm in that? I'll go and say, 'Good evening, I am Isabella Caccioppoli's daughter, Lorenzo's niece, I'm happy to meet you.'" She loves mathematics, and for her, and for many of her peers, he is a legend.

"Don't even dream of doing something like that. You don't know what he might say." Mum is afraid of this man; she is attracted to him and at the same time she is afraid of him. A fear that she was to overcome, but I only learned this much later.

Fig. 1.1 Isabella Caccioppoli

2

Chiaia
Winter 1958

"*Professore*, forgive me if I disturb you, but my daughter (and she gives the name) is one of your students. How's she doing? Is she good?" a lady swathed in a black and curly astrakhan fur querulously inquires as she approaches the table where Renato eats, or rather fasts, usually alone or with a few friends.

It seems that he looked up from his plate and that, with a look of derision—a look that, instead, those who knew him well describe as melancholic and even velvety—he replied, in his Italian made slower and a little slurred by the Neapolitan cadence: "Madam, I suggest you send the girl to work the via Domiziana and not to university. It's a waste of time..."

Naples is used to these excesses, paradoxically it welcomes them. His quips, caustic to the point of cruelty, are reported with a shiver of horror, but also with surprising benevolence. "The professor," they say in bourgeois salons, in PCI circles, in the Gradoni di Chiaia neighbourhood where he spends hours curled up drinking to the first light of dawn, "the professor is a genius." And stories about him abound.

Renato is a son of the city, in whose veins runs—so the legend would have it—the exotic Russian blood of the founder of the anarchist movement Mikhail Bakunin, and his oddities would therefore be the fruit of this singular, mixture of Slav and Neapolitan, at the origin of a long-lasting existential malaise the cause of which would be sought for a long time.

He lives in the Chiaia-San Ferdinando district, in that small world tucked inside the perimeter delimited by piazza Amedeo and Piazza Plebiscito, where aristocratic, bourgeois and proletarian, conservative and anarchist, intellectual and ignorant humanity all coexist. Every morning he leaves home on foot and crosses the invisible border that separates that circle closed off from the rest of

L. Foschini, *The Friction of Life*, https://doi.org/10.1007/978-3-031-65262-2_2

the teeming Neapolitan universe. Always at the same time, a quarter to eight, except on Sundays, he goes up via Chiaia, in summer more slowly, stopping every now and then as if finishing a thought. He continues along via Roma up to the corner with via Diaz and then down towards piazza della Borsa and, along the Rettifilo, to the university.

It is said that at night he goes into the Quartieri Spagnoli, that he crosses the ancient streets known as the "decumani", and that he drinks incessantly in the dark bars that open onto the alleyways populated until dawn by disreputable figures, but harmless to him. With them he talks and guzzles so that can't feel the furious intolerance that the obtuse bourgeoisie, "the living dead", always triggers in him because they are forever engaged in that idle and banal chatter that irritates him to the point of provoking unstoppable verbal violence, because stupidity hurts him to the point of wounding, as happens, moreover, with the students who stammer in his presence mathematical formulae whose meaning they don't seem to grasp.

Very often on Sunday morning, after the screenings at the cinema club, he goes to the PCI branch in Chiaia Vetriera, an office that faithfully reflects the neighbourhood in which it is located, elegant and posh, plebeian and cultured, next to the cinema Delle Palme, a few metres from via dei Mille. An elite branch that includes senators and deputies such as Mario Palermo and Vincenzo La Rocca, the editor of *l'Unità*, Mario Sansone, Luigi Marino and his girlfriend Paolina, daughter of the analytical geometry professor Maria Del Re. And party leaders, students and intellectuals. Caccioppoli arrives surrounded by a swarm of university students who he torments with aphorisms and mordant witticisms. He is a great talker, Caccioppoli, a magnetic conversationalist who loves to talk for hours and hours, tireless but always restless, always "possessed" almost as if from one moment to the next the subtle embroidery of connections should break. He enjoys reciting Rimbaud and relating the salacious *bons mots*, so congenial to his spirit, of Alfred Jarry and Eugene Ionesco. Everyone reveres him for his undoubted charm, the effortlessly displayed intelligence and eccentricity that emanates from his forebears.

3

Renato's "Grandfather"
1869

My grandmother, who has information that has always proven to be correct about him, perhaps gleaned from some cousin of his large family, states with absolute certainty that the much-rumoured maternal grandfather, Mikhail, was not his grandfather at all. Nevertheless for years Neapolitans seemed not to want to see the truth, clinging against all reasonable doubt to this fanciful ancestry, perhaps to give more weight to the brilliant mathematician's strangeness, to justify his charisma, to enhance his character.

Thanks to the patient and tenacious research of a professor of chemistry from the università Federico II, Carmine Colella, we have the recent discovery, in the archives of the International Institute of social history of Amsterdam, of a letter that has the last word on this matter.

On December 16, 1869 Mikhail wrote to the poet Nikolai Ogaryov from Locarno: "Dear friend, I wish once and for all to explain to you my relationship with Antosha and her actual husband." And he tells him what happened. Four years earlier in 1865—when he settled in Naples, convinced that it was one of the most favourable places to prepare the revolution—while he was conspiring, his young and beautiful wife, Antonia, was amusing herself with one of his most faithful followers, Carlo Gambuzzi, a Neapolitan lawyer and anarchist. In 1867 Bakunin moved with Antonia to Switzerland. Shortly afterwards the young woman confessed to her husband about her affair in the shadow of Vesuvius. Mikhail, understanding, set her free, and she chose to remain. But at the Lugano congress Antosha saw Gambuzzi again, and the passion exploded stronger than before. She gets pregnant. As in a nineteenth-century feuilleton, she hides her pregnancy from her husband who is always distracted by politics and, with the excuse of taking a trip, she gives birth in a

© The Author(s), under exclusive license to Springer Nature Switzerland AG 2024
L. Foschini, *The Friction of Life*, https://doi.org/10.1007/978-3-031-65262-2_3

village near Vevey. Gambuzzi arrives and takes the child with him to Naples. In the letter to Ogaryov Bakunin confides: "As for me, I suspected nothing. A year ago, in October 1968, an episode revealed everything to me."

Once again, the anarchist invites the woman to decide freely whether to stay or go to live with her lover. And once again Antonia chooses to stay. But, having gone to Italy to see her child, she returns to Lugano pregnant once more. At this point Gambuzzi proposes she give birth to her second child in Naples and leave it with him, renouncing it forever; she, desperate, begs Mikhail to recognize the two children as his. Bakunin accepts, but on condition that his wife terminates her affair with the Neapolitan lawyer once and for all.

Antonia gives her word. Mikhail believes her and confidently communicates to Ogaryov: "Their love affair is over." And he adopts Gambuzzi's children. Bakunin concludes the letter: "Antosha and I will remain together until the revolution calls me. Whereupon I will belong only to her and to myself." But what happens instead is entirely predictable. The affair between Antonia and Gambuzzi continues. After the birth of Carlo and Giulia Sofia, Renato's future mother, she was to give birth to another girl, Maria, who Mikhail recognizes. When finally Bakunin dies, Antosha marries her lover.

Over the years, Carlo, Sofia and Maria were to hear rumours that indicate Gambuzzi as their real father, but they always rejected them with disdain and tenacious firmness.

4

Salon Offritelli
1923

Villa Majo is a neoclassical house. It stands on a small plateau at the beginning of Via dell'Infrascata, the old name of the current Via Salvator Rosa. A road that once from the Vomero hill went down to the centre of Naples. Built at the end of the eighteenth century by Marchese Genzano, despite neglect and speculation, of the former beauty of the villa there still survive some flat columns surmounted by capitals and a small garden that boasts an enchanting little temple surrounded by trees. From the windows you can enjoy the breathtaking view of the gulf of Naples and Vesuvius. Now broken up into many apartments, the new occupants are perhaps unaware that within its walls Gaetano Donizetti, the native of Bergamo and the "most Neapolitan of musicians" lived and composed Lucia di Lammermoor.

In the 1920s, music continued to resonate in the rooms of the villa thanks to an old pianist of great talent. Clotilde Offritelli organized in her salon concerts for a small group of enthusiasts, who listened in religious silence to the performance of a symphonic masterpiece transcribed for piano four hands, or a string quartet of some sonatas for violin. Every Thursday, Renato Caccioppoli climbs up the Infrascata on foot to the villa, in whose living room he usually plays the piano.

Gianfranco Cimmino, who would later become a talented mathematician, is 15 years old and still attending high school when he hears him for the first time. He is won over by this engineering student who is only 4 years older than him and plays pieces by Brahms and Debussy divinely, and ever since he never misses a performance. They start seeing each other. A friendship is born. One day Renato invites him to viale Calascione, where he lives with his parents: "I will have you listen to something extraordinary," he promises him.

And that evening, for his friend, he performs on the piano the third act of Wagner's *Tristan and Isolde*, singing and playing, from the initial languid sound of the horn reporting that the ship carrying Isolde is not yet in sight, down to the last poignant lament of the desperate lover as she is dying. For Cimmino that river of melody is an impressive revelation, but not the only one.

The two youngsters spend their evenings discussing music until late at night. Caccioppoli describes only in words, but with astonishing precision, note for note, Strauss's symphonic poem *Death and Transfiguration*. Cimmino listens to him, rapt. In those meetings Renato talks about music and mathematics, but also of all the other passions that animate him to the point of torment: Proust, Rimbaud, whose poems he recites from memory, Pascal and his brilliant distinction between *esprit de géométrie* and *esprit de finesse*, down to cinema, another of his great loves. Over time he was to especially appreciate the films of the French masters: Duvivier, Renoir, Carné. But also Chaplin, Eisenstein, Rossellini.

After finishing high school in 1923, Gianfranco enrols in mathematics and Renato abandons his engineering studies and moves on to the third year of that same discipline. It is 1925. The well-known mathematician Mauro Picone is called on to teach higher analysis at the University of Naples. Renato is struggling with a degree thesis on Pfaffian systems, but he is in crisis. He would like to pursue a career as a pianist or conductor. Benedetto Croce, a family friend, suggests to him: "Caccioppoli, continue with maths. It requires a method that passion cannot give. Music will survive."

But the young man is conflicted. He starts attending Picone's classes. The professor, who is developing the theory of the Lebesgue integral, one day assigns his students the task of finding an example by virtue of which "a hypothesis formulated for a certain theorem were to prove essential." At the end of the lesson, the mathematician leaves the lecture hall and heads towards the exit when he is joined by a breathless young man, dishevelled, "scruffily dressed", who tells him, stammering, that he has found the requested example. It's Renato. "I invited him to come to my office," recalls the professor, "and he showed me a very elegant example that completely solved the problem… After some time, the roles were reversed and he became the master and I the pupil."

Renato, just graduated, is appointed Picone's assistant. A partnership, theirs, that was to last intact over the years. It seems as if we can see them both walking along the wide corridors of the faculty of Mathematics. An odd couple. The older one, 40, elegant, severe, with a round face on which his gold-framed glasses stand out. Next to him a young man half his age, dressed untidily, thin, snappy, always on the move like quicksilver (Fig. 4.1).

Fig. 4.1 Renato Caccioppoli

The master, a fascist from the start as he likes to call himself, and the anarchist, rebellious student, who after the fall of the regime, however, was to defend his professor thereby proving that he, beyond any summary assessment, while so intolerant of dictatorship and a victim of it, is capable of looking deeply within men and things.

5

Et velum scissum est
1925–1930

Since childhood, Renato was afflicted by constant inner torment.

The family does not understand the cause and worries about this child who amazes everyone with his intelligence, made even more prodigious by an excessive sensitivity. But it is when his adolescence is behind him, having reached the threshold of 20 years old, that something mysterious happens in him. A veil is torn in his mind, opening boundless expanses of knowledge before him.

In a short time, the very young mathematician attains surprising goals in his research and suddenly finds himself living in a state of exalted fervour. In the space of just 5 years, he publishes around 30 works which take him to a professorship at only 26. They are studies aimed at linear functional analysis and the theory of functions of real variables, with particular attention to integration, ensemble functions, quadrature of surfaces and linear approximation. In a note from 1926, anticipating in a particular case the well-known Hahn-Banach theorem, he demonstrates that, through successive passages of the limit, a continuous linear functional, defined initially only on the set of all continuous functions, can be extended over the set of all limited functions of Baire. The ideas that he conceives at just 23 in the study of surface squaring also contain the germ of the profound theory of measurement and integration in dimensionally oriented sets: "I have outlined," he writes in those days, "in its essential features, a new chapter of modern integral calculus." And again, in a note of 1928, for the study of the passage of the limit under the integral sign in very general conditions, the young Caccioppoli introduces the important notion of the "family of functions uniformly with limited variation".

© The Author(s), under exclusive license to Springer Nature Switzerland AG 2024
L. Foschini, *The Friction of Life*, https://doi.org/10.1007/978-3-031-65262-2_5

I know, these words mean very little to many of us, but to know Caccioppoli it is necessary to understand the decisive influence he had on the development of mathematical analysis and on an entire generation of Italian analysts in a period, the 20 years of the fascist regime, in which Italy was isolated from the rest of the world.

Renato writes his works in one go, as if to clarify first of all to himself and to those who are able to follow his arguments the terms of the problems. In his notes he leaves an often concise, succinct outline of the perceptions that led him to the solution. His works are often difficult to read but, for those who want to and are able to understand them, they often offer the intoxicating sensation of participating—alongside a refined intellect—in the construction of a new chapter in science.

However, if it is difficult even for many mathematicians to grasp the meaning of his research at first sight, it is even more complicated to grasp what happens in him starting from this moment. It is as if the flashes of sudden intuition, fixed on paper in a few synthetic lines, were followed by the darkness of an anguished solitude, because the torn veil that led to self-awareness, arousing in him an immense happiness, immediately afterwards opens up inevitable abysses of terrifying emptiness.

Leonardo Sciascia wrote something so profound about Ettore Majorana, the famous Sicilian physicist, as to throw light on the Neapolitan mathematician, too: "In the precocious genius [...] life has a sort of insurmountable measure: of time, of work. A measure as if assigned as indefeasible. In the work, as soon as a completeness, a perfection has been attained; as soon as a secret has been completely revealed, as soon as a perfect form, that is to say, a revelation, has been given to a mystery—in the order of knowledge or, to put it approximately, of beauty: in science or in literature or in art—immediately afterwards comes death. And since it is as 'one' with nature, as 'one' with life, and nature and life are as 'one' with the mind, the precocious genius knows this without knowing it. For him, doing is imbued with this premonition, this fear. He plays with time, with his time, with his years, in deceptions and delays. He tries to extend the measure, to shift the boundary. He tries to avoid the work, the work that once completed is concluded. The work that ends his life.

And in fact, dejection and anxiety, similar to a grip that tightens and oppresses the chest, are unbearable sensations that Renato has known well since childhood, but which gradually become increasingly harder to manage. Of course, he can get a break from this unspeakable torment, or rather pour out his pain in the performance of sublime musical pieces, in the prodigious notes where equally brilliant minds have poured out similar suffering. But then the dizzying sensation of nothingness returns to invade his soul. It is at this point in his life that he begins to drink. And there is no point in finding reductive medical, sociological or psychoanalytic interpretations to understand the reason why. For Caccioppoli, alcohol is an expansion of the self.

Fabrizia Ramondino, who knew him and had investigated his life has a good explanation for this: "Just as the expansion and contraction of the heart correspond, metaphorically, to the relaxation of all the limbs after having been in chains, a deeper inhalation of breath after being locked in a narrow prison, so if we are gripped by anguish, we expand in exhilaration. Alcohol is not just a vasodilator—it is a soul-dilator." As it was for Baudelaire, intoxication and oppression are recognizable in the concentration and expansion of space and time. Space and time become profound: as Roberto Calasso wrote, they nullify "their appearance of opaque surfaces and reveal, little by little, a succession of [theatre] wings between which one gets lost".

To consider Caccioppoli's work while removing his existential restlessness, which from this moment on profoundly marks his scientific career, would be an unpardonable error.

6

Salon Del Re
1930

Houses contain memories, and more. This time, unlike Clotilde Offritelli's salon, this apartment exists still, preciously intact after almost a century.

Professor Maria Del Re's home is located in via Carlo Poerio, on the corner with Piazza dei Martiri, in that quadrilateral of alleys and streets where many of the people who are part of my story live their lives.

As I enter the dark hallway, coming up the worn stairs, getting into the elevator where I put a coin in the box, like when I was a child, to set it in motion, I get the feeling of being in one of those places, very rare now, where it seems that—by virtue of a crystallization process that changes a substance from the fluid to the solid state—the flow of time has stopped, as if fixed in far-off days and hours, even though today the apartment is occupied by the young children, one of whom is called Renato, of Senator Luigi Marino, who accompanies me.

I look around. In the living room there is a sofa upholstered in red damask with a back edged in mahogany, the same as the dining table and consoles; the nineteenth-century paintings have gilded frames. On the tables and walls, many old photographs. The piano where Caccioppoli played on the long evenings spent in this house stands against one wall.

Wandering through the rooms and listening to Marino, I discover that Maria Del Re was a very particular woman about whom too little is known, unfairly, but above all she was a very important person in Renato's life. Having come to Naples from Reggio Calabria, Maria graduated in 1922. Assistant to Pasquale Del Pezzo in the projective geometry lab, she was to be the first female tenured professor in a mathematical discipline. I look at a photo of her together with Caccioppoli: the round face, the ample forehead, the cheerful, lively and intelligent look, the mouth open in a smile that reveals beautiful, very white teeth and black hair gathered in a bun at the back of her neck, as peasant women used to do (Fig. 6.1). Next to her the young mathematician, who is 10 years younger, looks like a boy, with his handsome, sunken face and pronounced cheekbones, illuminated by a hint of a smile and a look that is not yet melancholic and suffering as in the years of maturity, but crossed by a shadow of restlessness so profound that I am struck by it.

Renato, who has perhaps more than a simple friendship with the professor, loves to spend the evenings in via Carlo Poerio Renato, often stopping for dinner, playing chess with the hostess, delighting the guests with his brilliant observations and his dazzling bons mots, playing with the cats, and conversing with Maria's young niece, Michelina Tommasini affectionately known as Nella, 6 years younger than him. This girl with amber skin is the natural daughter of Maria's cousin and an Eritrean. The professor's goddaughter, Lina Maione, also lives in the house. Other mathematicians were to be present over the years: from Gianfranco Cimmino to Giuseppe Scorza Dragoni and, going back in time, Professor Del Re's old teacher, Pasquale Del Pezzo.

I leave the house clutching a pack of letters from Renato to Maria. They are also precious because, like all the Caccioppolis starting with my grandfather, he never liked to carry on correspondence except for reasons of work. They are letters exchanged mainly in the period in which the young professor was called to Padua, in 1931, at just 26 years old, to occupy the chair of Algebraic analysis, and they continue until 1934, on his return to Naples. On reading them I discover, behind the proverbial confidentiality, aspects of his character that were hitherto unknown and had eluded the stereotypical portrait made of him over the years. Humorous, witty, critical, but also with traits of unexpected sweetness and an intact childish playfulness.

Fig. 6.1 Maria Del Re and Renato Caccioppoli

7

Padua
1931–1934

Renato arrives in Padua immediately after the Christmas holidays. He is surprisingly cheerful. Not at all upset at the idea of leaving his home for the first time, breaking the delicate balance of his habits. He goes down to the Leon Bianco hotel, where other professors stay and show themselves to be friendly, even affectionate towards him. He begins to hold his seminar, followed by a large group of students who participate with enthusiasm. He is serene, reassured by the atmosphere that surrounds and inspires him; in his first letters to Maria he talks of tranquillity, ease, comfort. He even goes so far as to have fun, he who was so reluctant to take part in any type of sporting entertainment, going on trips to the Colli Euganei perched on the crossbar of the young mathematician Pietro Pagani's bicycle: "An iron bar," he writes, "has harshly ruled the less conspicuous part of my person (*buium ammaccavit*, Nella would say)." During the long bike rides, his colleague, whom Renato affectionately calls Pichetto, confides in him. He is sad in those days—he tells him—"because his wife can't decide to have a child".

And Caccioppoli perfidiously comments: "But why on earth do those people have to bring unfortunates into the world at risk of their becoming stunted, idiots, professors of algebraic analysis and so on?" In this state of mind, so unsuitable to his restless temperament, he lets his imagination run wild (Fig. 7.1). He writes zany nursery rhymes that recall *Ingarrichiana*, a genre of bizarre and absurd poetry invented in the first half of the nineteenth century by a Neapolitan judge at the time when Naples was under Bourbon rule, namely one Ferdinando Ingarrica, and which was to enjoy lasting fortune, influencing futurist poets and actors such as Petrolini. The intended

Fig. 7.1 Renato Caccioppoli

victims of his mockery were his mathematician friends, the Neapolitans Gianfranco Cimmino and Giuseppe Scorza Dragoni.

These bizarre poems are often accompanied by little letters that he sends to the two very young girls, Lina and Nella, who live with Professor Del Re. Renato has invented a childish language for them, made up of distorted words identical to those used by children as they are learning to speak, which made a big impression on me because it reminds me of what Marcel Proust used to do in the letters he wrote to his lover Reynaldo Hahn to dispel his depression and entertain him. It is a sign that Renato too, in this first period in Padua, feels the need to rediscover a glimmer of lost childhood. A space in which his tormented nature can take refuge and look for a break in the game, in a heart-breaking desire for happiness.

Dear Bwutussa, fankoo fo ve letta, witch weachumme attan oppotchoon moment, when I wuz feelin dark and meejetly blighted up. Aydoo everything as oo hav awainjed, and eye eatalot. But I never go doodoo, and this is a seeryus inconvenyens at give me umpiteen reason to regwet Maliassa (Maria Del Re). Nor do I falldown on ve street, those eevnings I walk awound in ve dark and ve fog, so thik thatchoo woodn't see the tip of oo little nose. But I alwaze get home by minnight—and only on theez days will I take a liddle leaf (fo wun o'clock) and go to ve Opewa.[1]

Play, but also affection and protective feelings: "How did Nella's exams go?" he asks Del Re. "Has she started her degree thesis?"

Since he was a child, Renato has been surrounded by women. His mother, his aunt and then Maria and her protégées occupy a central place in his young life, but he, as if to defend himself, hides the fascination he feels for the female universe behind ferocious, ironically contemptuous jokes. This is how he tells Maria about his days, which follow one another full of boredom: "That deeply disheartening thing called 'the natural order' which brings our life closer to a problem of analytical mechanics with pre-established initial data, and ensures that methods similar to those of Volterra in that famous memoir on mathematical biology can be applied to the study of the social animal, called big fish and little fish. The funny thing is that I, being a little fish, have to pretend to be a big fish and puff myself up like an eel... in the presence of the first-year cod and the third-year salt cod (women above all)." After the first few months, the atmosphere becomes increasingly heavier. He feels the growing burden of fascism weighing on everything and this is followed by an uncontrollable intolerance, an irritation, in the face of mediocrity, and disgust for the servility of those around him. A mere glance, a brief conversation with a colleague, or an exchange of words with a professor, also staying at the Leon Bianco, and he finds himself faced with the same justifications, the same indulgences in the face of superficiality, narrow-mindedness, the drift into everyday paltriness, conformism. Everything outrages him. The flood of vacuity and vulgarity.

He wanders for hours alone in the city that he feels is extraneous to his suffering, walking the misty streets under a thin and uninterrupted rain. For him, Padua is Cimmeria, the imaginary antediluvian region sung by Homer, populated by barbarians, where the land is barren and harsh and shrouded in mist. He walks with his eyes lowered, mumbling incomprehensible words to

---

[1] Dear Blutussa (Nella, Maria Del Re's niece), thank you for the letter, which came to me just at an opportune moment, when I was feeling dark and I immediately brightened up. I do everything as you have arranged, and I eat a lot. But I never poop, and this is a serious inconvenience that gives me the umpteenth reason to regret Maliassa (Maria Del Re). Nor do I fall on the street, those evenings that I walk around in the dark and in the fog, so thick that you wouldn't see the tip of your little nose. But I always get home by midnight—and only on these days will I take a little leave (for one o'clock) and go to the Opera.

himself which are none other than the verses of the Odyssey: "There the people of the Cimmerians dwell, enshrouded in mist and darkness …"

He is spied on. A circular from the prefecture of Padua, run in those days by the political police, informs us in the usual ungrammatical language that "Professor Caccioppoli Renato [...] seems to be of Russian origin, and even if this would not have given rise to criticism for his political conduct, however, it is clear that he is not a member of the Party."

After 2 years the weight of the regime also begins to affect his work, influencing him heavily. The bitterness and nausea that seize him shine through in the words that in November 1933 he wrote to Professor Giulio Andreoli, a fervent fascist. It is a letter which, beneath the apparent irony, does not hide his bitter observations. Called upon by his colleague "to observe ministerial regulations", Caccioppoli assures him that "no one appreciates their superiority more than me", but immediately afterwards he abandons sarcasm to affirm his loyalty to his principles "so well known to my friends that no one writes to me anymore."

8

Inert as Life Is
1933

Suddenly the correspondence between Padua and the Del Re "family" ceases. It is perhaps precisely in this period that a mysterious event occurs: Caccioppoli suddenly disappears. All trace of him is lost for days. I looked for evidence of what happened in my visits to the State Archives, but in vain. Yet this episode that is so disturbing and revealing of Renato's personality certainly took place, and my grandmother's stories provide me with a small detail.

One night, Lorenzo Caccioppoli's family is awakened by the telephone ringing. Rosa replies. On the other line is one of her cousins. He has received— he says—a request for help from Peppino Caccioppoli. The surgeon is in a state of anxiety bordering on desperation. His son, who teaches in Padua, has disappeared. The family hasn't heard from him for days. My grandmother is begged to contact one of the more questionable members of the clan, whose degree of kinship is lost in the meanders of this very large family and so was very hard to reconstruct. This is Saverio Polito, an active member of the fascist secret police, the OVRA. My grandfather, in his modest emotionality, must certainly have felt greatly disturbed at running to the aid of that much revered uncle by whom he had been repudiated 30 years earlier. But it must also have been very difficult for him, an anti-fascist, to communicate with a relative he had always kept at arm's length for his political ideas. Yet he doesn't hesitate to send him a message. Polito is alerted.

The young mathematics professor is found sleeping on a bench in a railway station in Milan and is arrested for begging. He had removed all money from his pockets, had grown a beard and, dressed in rags and dirty, had boarded a third-class carriage on a train bound for Milan. The police describe him as in a confusional state. When asked why he was there, like a tramp, he apparently

L. Foschini, *The Friction of Life*, https://doi.org/10.1007/978-3-031-65262-2_8

replied that he wanted to see how desperate people lived without a home. Five days have passed since he disappeared.

It was not to be the last time. In Naples too, in the years to come, Renato—they say—would sometimes crouch at the corner of the alleys of the Quartieri Spagnoli to ask for alms.

Perhaps it is precisely after this mysterious absence that he resumes his correspondence with his distant friend: "Dear Mariassa, the long suspension of our relationship gives rise in me the desire to resume it, if there is still time, since in you the urge to continue that suspension has probably become definitively acute. But perhaps I am slandering you… The fact is that something worthy of being communicated must have happened by now, 'no matter how inert life may be.'" So, Renato's pen lets slip this adjective, which defines what existence is for him: stagnant, devoid of movements, devoid of heart, devoid of affection. Inert.

9

Homecoming
1934

"Dear Mariassa, after your postcard from Rome I was expecting a telegram," he wrote on 7 June 1934, "that would have announced the triumph of our girls, classified, examined, sounded out, complimented and asked for in marriage by the members of the commission. I take this opportunity to inform you that I feel no worse and that my soul is no more in peril than usual. Appetite, nutrition, replacement parts, hormonal circulation normal, at least for me. Conquests are also stable. I hope that this good news will come to comfort you in my absence, which is destined to be prolonged. Our ineffable Minister, so worthy of those who command (and those who command him), has exempted us in some way from the lessons, as long as our presence ensures the continuity of 'academic life'. And so, among the ministerial circulars, the redoubled zeal of our vacillating Rector, the fezzes (no pun intended) of the Academic Senate, the GUF, the Fascist University Groups, sitting in the great hall and the balilla who pis... in the courtyard, the university has become an authentic barracks worthy of Courteline (an allusion to *Lo squadrone si diverte* by Georges Courteline, author's note), [...] and yet there isn't too much to complain about... When I return permanently to Naples I will feel as if discharged, and I will send you one of those beautiful postcards in superimposed colour, which can be admired in the tobacconists' displays. In the meantime, I'm going back and forth to Venice where the Biennale offers new and interesting documentation of that universal mediocrity which has been euphemistically called 'clarification'. Let me have your news and that of the *jeunes filles en fleurs*."

Renato loves Proust, he shares his neuroses, his impatience, his complicated sensibility, his difficult amorous mechanism, but above all his morbid

L. Foschini, *The Friction of Life*, https://doi.org/10.1007/978-3-031-65262-2_9

attachment to his own habits. Three years spent in Padua, far from his city, in addition to making him feel the pain of uprooting, have sharpened his tendency to see the world around him differently, a lucidity that quickly degenerates into disillusionment, if not even into desperation. Everything around him disgusts him. It is the moment of triumphant fascism, of dull enthusiasms, of vulgar exaltations. Now, called to the chair of Group Theory, he is preparing to return to Naples. He couldn't imagine himself anywhere else.

He is 30 years old. He has become the Renato Caccioppoli that the Neapolitans were to learn to know, whose image will be fixed forever in the memory of the city. Body of a skeletal thinness, unkempt beard on his sunken face, a mocking and sometimes nauseated look, barely veiled by the alcohol drunk at will. Aggression that erupts unexpectedly, flashes of tenderness, the cultured language that speaks of endless reading, Pascal, Rimbaud, Dostoevsky, mixed with the verve of slum vernacular. Deep and enigmatic, but also witty and a raconteur. Extraordinarily elegant, surprisingly childish: 'o Professore has become 'o Genio (Fig. 9.1).

Fig. 9.1 Renato Caccioppoli

10

Salon Benzoni
August 1937

To look for other traces of Renato I don't stray far from Naples. I just need to reach the Sorrento peninsula, where the Caccioppolis come from. Not only Joseph, but also his father Domenico, like him a surgeon of renowned skill. From Vico Equense where they were born, I move to Capo di Sorrento. Once again visiting a house that holds within its walls echoes of a meeting that remained memorable. The one between Renato Caccioppoli and André Gide.

La Rufola is a splendid villa, owned at the time by the Marchesa Giuliana Benzoni. It stands precipitous over the sea and overlooks the baths of Queen Giovanna, where from blue the water turns prodigiously green. Behind the now rusty gate I glimpse a long avenue, and in the background a beautiful yellow house and the garden where Benedetto Croce, Gaetano Salvemini, Giustino Fortunato and Alberto Moravia strolled in a very particular atmosphere in which "anti-fascism took on an air of holiday," the marchesa would recall years later, "and where even the rigorous, the committed, and the phlegmatic would loosen up." The villa is one of those places where by virtue of fortunate coincidences, politics, literature, and society converged and at one time, before his return to the Soviet Union, also the creativity and madness of a special neighbour, Maxim Gorky, with his picturesque colony of guests.

I have come this far to recall an evening so imprinted in the memories of some of the people who were present that traces of it can be found in their diaries and in some articles.

It is the end of August 1937. In Gide's honour, the marchesa Giuliana extended invitations to her Neapolitan friends inviting them to Sorrento. Welcomed to Rufola with all honours, the writer, in the company of the young translator and professor of philosophy, Robert Levesque, is taken to the

© The Author(s), under exclusive license to Springer Nature Switzerland AG 2024
L. Foschini, *The Friction of Life*, https://doi.org/10.1007/978-3-031-65262-2_10

library of the house, where his complete works are lined up in full view and where he signs the roll of honour of illustrious visitors.

In the last days of August, a light but fresh, almost cold breeze is already coming from the sea and the guests give up on the idea of eating on the splendid terrace overlooking the rocks lashed by the waves below. They eat in the dining room, a large, intimate yet austere room. On a long console table set against the wall stand bottles of wine, jugs of water and serving plates. The guests serve themselves and then sit haphazardly around a large table. As well as Giuliana there is her stepfather Carlo Ruffino, her mother Teresa's (Titina) second husband. There are Nicky Mariano, Bernard Berenson's secretary and lover, just arrived from i Tatti, the Florentine villa owned by the famous art critic, and Nicky's sister, Baroness Alda Anrep, custodian of the magnificent Tatti library of which her husband Egbert is administrator. And then Pietro La Via, marchese di Villarena, recently moved from Naples to the magnificent and ancient fourteenth-century residence in Massa Lubrense. A diplomat manqué due to his criticism of the regime, poet and philosopher, he corresponds with Gide but also Henri Bergson, Maurice Ravel, and Thomas Mann.

"We are waiting for a young friend who is joining us from Naples," the marchesa tells Gide. "He is a mathematician called Renato Caccioppoli, nephew of the anarchist Bakunin. A genius!" loudly proclaims La Via to give greater power to the quality of the arriving guest, then adding in a low voice with a sad sigh to Robert Levesque, sitting next to him: "But who knows if he will come."

At the table the conversation is heated. All convinced anti-fascists, they look at Gide waiting for a word from him to illuminate the dullness of Italian cultural life, but they are soon disappointed. Gide is still strongly shaken by the violent impression that his recent trip to the Soviet Union made on him and hesitates to wax indignant at fascism which, compared to what he has seen, strikes him as a rosewater dictatorship. Only 2 months before, in June, came the publication of *Afterthoughts on the USSR*, exactly 1 year after *Return from the USSR*, which caused so much scandal and uproar among communists not only in France, but throughout Europe. Levesque makes the situation worse, arousing a wave of indignant disappointment, by downplaying Mussolini's rise speaking of "*marche sur Rome en sleeping*".

But the guests insist, they don't want the writer to underestimate the drama their country is experiencing: "Even if it may appear to you in apparently more discreet forms," La Via explains to him, "in reality we feel the weight of this regime that is crushing us. You should know that fascism is an authoritarian system that erases all freedom as happens to all peoples dominated by dictatorship."

Levesque intervenes to help Gide and apologizes for the hasty judgement that seems to have disappointed those present: "Knowing what is happening in Russia we have had some hesitation about getting indignant about fascism but," he justifies, "now we understand clearly that the spirit is crushed in Italy as it is beneath other dictatorships." Conciliatory words, which alleviate the tension and put a peaceful end to dinner.

The hostess stands up, inviting the guests to follow her to the living room that overlooks the garden. The ladies barely speak, except for Alda Anrep who, part Russian, is fired with a barely restrained exhilaration, and she confides with an inspired air to those next to her that she intends to question Gide on the "supreme issues." Marchesa Benzoni is silent. Pietro La Via is also anxious to hear what Gide thinks. He has known him for years, at first in Paris, and they are united by common passions including, last but not least, a strong predilection for beautiful young men. The "marchesino", as Levesque calls him in his diary due to his stature, was the life and soul - as long as it was possible—of the "Posillipo club" frequented by the most cultured and liberal people in the city. Even though La Via is considered an anti-Crocian, Don Benedetto was often present at those cultivated and refined meetings. But now, for his ideas, La Via was forced to exile himself in Massa Lubrense.

This is the air you breathe in the spacious room with high plastered ceilings. An atmosphere full of anxiety and questions that the famous intellectual from France seems reluctant to satisfy all that much, and while he always responds with laconic courtesy, he appears slightly bored.

It's already very late when, from the large French window wide open onto the terrace, a figure so slender as to appear stylized, composed of a few geometrical lines, like a sculpture by Giacometti, enters with hesitant step. Gide sees an emaciated young man, who strikes him as being little more than 20 years old, come forward.

"I took my courage in both hands," the newcomer says in one breath, heading towards the hostess and greeting her warmly, as if coming from Naples to Sorrento for him was an undertaking beyond his strength. La Via, observing him, has the impression that, despite the very fragile appearance, he is completely at ease in the fleshy shell of that evanescent body.

The conversation resumes, after the introductions and pleasantries, but is no longer as lively as before. The talk is still of war, of politics, little or nothing of literature, however the words seem to fall on deaf ears. Guests notice that Gide's attention is elsewhere.

André only has eyes for Caccioppoli. He surreptitiously observes that sunken face with the prominent cheekbones, which despite being so beautiful—he will later write—reminds him of Leopardi, a copy of whose death

mask he has on his desk in Paris. He finds there the same expression of profound sadness, infinite bitterness, a certain impassioned desolation that moves
him even more because he considers him a boy, even though he was introduced to him as a professor of analysis at the University of Naples (Fig. 10.1).

"May I ask how old you are?" he asks him, and he is stunned to learn that
he is 34. Renato smiles at his amazement and tells him that a dispensation was
required so that he could take up the position at the age of 26.

Made aware of the key topic of the evening, the Soviet dictatorship and
fascism, the young mathematician gives Gide the impression that he still has
faith in Stalinism, although he limits himself to expressing this conviction
with two spare remarks, and then closes himself in a strict silence from which
he seems to wish to emerge no longer. The writer does not give up. He wants
to find out the reason for the doleful anxiety he sees on his face, and delicately
tries to establish a more direct contact with him.

"Do you have difficulty," he asks, "appearing even younger than you are,
when it comes to relationships with your students?" Renato, who up until
then had responded in monosyllables, begins to speak with unexpected
vibrancy: "The students? An unbridgeable gulf has opened up between us. No
contact is possible anymore. I find myself talking to deaf people who have ears
only for slogans. 'Believe obey fight' are the watchwords that exempt them

Fig. 10.1 Renato Caccioppoli

from any reflection, from any search, from any effort of the mind. With no curiosity anymore, they have delegated others to think for them, and they repose in idleness. Yes, fascism has made these young people lazy. Intellectually, of course, because they only exercise their bodies, bodies without brains."

Robert Levesque is struck by Renato's ardour and desperation. He has the impression that on those slender shoulders the burden of the tragedy of his country weighs heavy. Gide is silent, he does not want to close the breach that he has managed to open in the soul of that enchanting and tormented character like a hero out of Tolstoy or Dostoevsky who, with "eyes sparkling with intelligence", meanwhile continues without stopping: "What will Italy become with young people formed like this, or rather deformed like this?"

Levesque, seeing that emaciated face come to life again, his black eyes shining with an intense light, is struck by a profound emotion and asks something to the marchese La Via, who is sitting next to him and who answers him almost in a whisper so as not to make himself heard: "Renato is an extraordinary person. His mood, his melancholy which often overflows into desperation, keeps his family and all of us who love him in a state of alarm. There is always a fear that at any moment he might make a fatal gesture." And, as if that wasn't explicit enough, he specifies in a breath: "Suicide."

"What loneliness is ours?" in the meantime Renato continues to ask himself in his exquisite French rendered even more melodious by a marked Neapolitan accent. "We have no one to turn to. We are truly, completely abandoned to ourselves."

"You have Croce," Gide reminds him.

"Of course, Croce. We know his position. His disapproval of the regime, but it is not easy to converse with him on these themes. He will tell you about his books, he will tell you a few anecdotes…"

"You're wrong, dear Renato," interrupts the hostess, "Croce is a shy man, but a young man who knows how to take him will be able to make him talk and will be able to get a lot from him."

This dialogue, which leaves those present speechless, must have been so intense and full of deep pain that, years later, some of the people present at that evening felt the need to remember it. Gide was to do so after the war in the *Journal* and above all in an article in *Arche*, Levesque was to do the same in his diary, and after Renato's death the marchese La Via did likewise in a recollection that appeared in *il Mattino* in which, as if to avoid hasty conclusions being drawn, Gide's homosexuality, like his, being well known, he hastened to add: "Don't think of anything in particular, do not burden yourself with analysis or psychoanalysis according to the custom or rather the obsession in vogue, but be satisfied with a faithful description. None of us were intelligent or idiotic enough. On the other hand, in that circle of friends it wasn't the words that counted and not even the ideas. What was striking was the unexpected and strong impression that Renato had made on Gide."

The 68-year-old writer and the little more than 30-year-old mathematician talk to each other until the small hours, as if the world around them had suddenly disappeared. They converse on the threshold of the terrace, secluded, regardless of the coolness and the humidity coming from the sea. What do they say to each other, why they smile sometimes and other times assume a serious, intense attitude, no one can know.

The evening is at an end. Gide stands up and saying goodbye to Pietro La Via whispers to him: "Ah! This man, so young. Thank you for introducing me to him. He is a soul."

11

State Archives: Rome
2020

Renato and André were to see each other again the next day and then other times over the years. I find traces of this in Rome, in the State Archive where they keep a tiny, slim folder on which there is written in italics with a red pencil only a surname: Caccioppoli.

I scan the few sheets eagerly. The pages I have before me span the years of Renato's life starting from his stay in Padua until his death. From a confidential registered letter from the Naples police headquarters to the political division of the Ministry of the Interior dated 1 April 1937, 3 years after his return from Padua, I discover that there was a report, according to which "Caccioppoli is allegedly spreading sensational news, commenting on it in a manner unfavourable to the policy of the regime. We have not, writes the police chief "found proof of these rumours, but the appropriate order for political surveillance has been reinstated concerning Prof. Caccioppoli, for which it is necessary to report any useful unexpected event in due time."

I also find the account of Gide's presence in Sorrento. The fascist secret services believe the writer is an extremely dangerous Bolshevik. No one informed them about the change in his political views, which got his comrades and European intellectuals so worked up on his return from the Soviet Union. And no one read what had been written a few weeks earlier in Pravda, the official organ of the Russian Communist Party: "Gide is the typical representative of a decadent bourgeoisie; he is an individualist." They are convinced, on the contrary, that this distinguished gentleman is a menacing communist, since the day after the *inoubliable* dinner at the house of Marchesa Benzoni, he breakfasted on the terrace of the Tramontano hotel with two gentlemen "who came from Naples to meet him", all highly suspicious.

L. Foschini, *The Friction of Life*, https://doi.org/10.1007/978-3-031-65262-2_11

Alerted by zealous French tourists who recognized their compatriot on holiday in Sorrento, the prefecture of Naples sends, in a language so convoluted as to require a more than twisted mind, a registered confidential note, addressed to the Ministry of the Interior and the political division of the police.

André Gide—French communist
October 1, 1937I inform this Ministry that the aforementioned foreigner with
  French passport No. 18419 stayed in Sorrento from 2 to 18 August as a guest
  in the Tramontano Hotel.

  He was accompanied by his compatriot prof. Levesque Robert born on...
passport N°...
  The aforementioned Gide—whose presence in Sorrento was noted by some of his
compatriots, by whom he would have been recognized for his writings—led an
almost solitary lifestyle during his short stay in said locality. Subsequent to investi-
gations ordered for this purpose, it turned out that the aforementioned Gide had
telephone communication with the Terme di Agnano, however it has not proved
possible to identify his interlocutor.
  The following day Gide received a visit in the hotel from two men coming from
this capital who stayed for breakfast with said foreigner, leaving the same day. The
investigations attempted to identify the two visitors have given a negative result.
  Prof. Renato Caccioppoli, son of Giuseppe, has been the subject of previous cor-
respondence with this Honourable Ministry: for this purpose, I refer to the paper
from the local Police Headquarters N° 1019339 of April 1st.
  It was not found that the aforementioned Caccioppoli went to Sorrento during
the period of Gide's stay there, but it cannot be ruled out that one of Gide's above-
mentioned visitors is Caccioppoli himself.
  I assure you that Gide, were he to return here, will be subject to careful, cautious
vigilance in order to follow activities and contacts and the Hon. Ministry will be
informed of any useful eventuality.

—THE PREFECT
(MARZIALI)

The document is dated October 1937, but with great surprise one morning, on returning to the State Archives, I discovered another note that I had previously missed. It's from the political police, and it's far shorter but much more disturbing; it dates back to a little more than a month before, only nine days after Gide left Sorrento. The note is very different from the first; and written in correct Italian by a person who seems to know, or rather report appropriately, the content of the conversation between Caccioppoli and Gide on that beautiful summer evening at Rufola or at breakfast the following day on the terrace of the Tramontano hotel.

POLITICAL POLICE DIVISION

Rome, 27 August 1937*A trustworthy source has reported that the French econo-mist André Gide—known anti-fascist—is currently in Italy. He allegedly had a meeting recently, in Sorrento, with Prof. Caccioppoli of the University of Naples. In this meeting, Gide allegedly entertained Caccioppoli with his own political beliefs, and our informant reports on this as follows: "Gide's central idea is the transience of the fascist phenomenon. He does not see in fascism a basis, a doctrine, a thought. He excludes any probability of Mussolini's succes-sion. He draws a parallel between Italy and Russia, concluding that fascist ter-rorism is clever, disguised, but worse perhaps than Russian terrorism."*

Caccioppoli attributes these words to Gide: "A regime, which rests only on a man's personal fortune, cannot be more than a transitional regime!"

—THE DIRECTOR
HEAD OF THE POLITICAL DIVISION

"Caccioppoli attributes these words to Gide…" To whom, in his entourage, had Renato confidently told the content of the conversation? Which friend or acquaintance had collected his confidences and made sure to report them punctually, in cultured and appropriate Italian, so different from the crude and contorted reports of an obscure official in the secret services? Or, an even more disturbing hypothesis, was it someone who was present at Rufola that evening who intercepted the conversation?

12

Taking the Cockerel for a Walk
1936

So Renato is monitored, spied on, and he knows it. This makes his relationship with the outside world even more difficult. He becomes gloomy, withdraws into himself, and confides to his friends: "Naples is a swamp, and we are the sick fauna of this swamp. Cowardice makes us fat and kills us at the same time."

Black shirts move around him, vulgar slogans resound, absurd provisions are issued. Nevertheless, when it seems that everything should lead him to isolate himself, so as not to be touched by the horrors that surround him, suddenly an irresistible call awakens in him, a vital and childish spark that leads him to acts that, rather than rebellion, appear surreal, even futuristic. This is the case of the cockerel on a lead.

A regime directive prohibits men from walking with small dogs because it is a symptom of not very virile attitudes, in contrast with the exaltation of the fascist male. Only a few years earlier Curzio Malaparte wrote, with such an exaggerated enthusiasm for the Duce that it verges on mockery: "The sun comes out / the cock crows / or Mussolini mounts his horse…" And here is Renato who one morning is coming up via Chiaia as usual, but walking at a slow, rhythmic pace because he actually has a cockerel on a leash. The presence of the poor bird taken for a walk in one of the most elegant streets of the city arouses the laughter of passers-by and ridicules the fascist myth of masculinity. I found confirmation of this episode in a medical record that I shall be discussing later.

Many of his extravagances are the stuff of fabulous stories, but there is a completely unpublished one told to me by a relative of my grandmother, over 90 years old, and which frankly strikes me as delightful. Having arrived in

L. Foschini, *The Friction of Life*, https://doi.org/10.1007/978-3-031-65262-2_12

Rome in '37, Tonino—this is his name—is invited by a cousin to visit a friend, a distant family member: "You absolutely must meet him," he says. "He is a special character, a mathematician. Apparently, he is a genius."

The two fellows set off on foot up Via Capo le Case and come to a strange building with closed shutters. The young man, who comes from the provinces, discovers to his great surprise that he finds himself in a brothel. He and his cousin go up the stairs, walk through a red room with a wealth of golden stucco and are introduced into a bare room with a bed in one corner and, in the centre, a small table behind which sits a man who seems only a little older than they. The strange and seductive character stands up and greets them amiably, and when asked why he is there he replies smiling: "It's a very pleasant place, I come here often when I'm in Rome because it's the best place to study."

13

State Archives: Rome
2020

While I check to see if there is still something to read in the slim pile of carbon copies, two extremely light pieces of paper slip from the folder, brush the wooden tabletop, and fall onto the floor. Bending down to pick them up, I see the header and widen my eyes in surprise as I focus on the name: *Rosa Caccioppoli* (Fig. 13.1).

Grabbing them, I wonder how my grandmother had ended up in dispatches of the fascist secret police concerning Renato. Out of curiosity, I scroll through them avidly, even finding, unexpectedly, my father's name.

Fig. 13.1 Rosa Caccioppoli

Rome, 15 December 1940, year XIXMrs. Rosa Caccioppoli said that the Italian People have now lost faith in the Duce, because the continuous losses suffered by Italy are the index of the disintegration and the imminent decline of fascism which was believed to be enduring and which actually lasted too long. That this Regime will disappear as all the others have disappeared, to make room for new ideas and new, less exhausted men.

Caccioppoli said she had learned from her son-in-law, Nicola Foschini resident in Naples and part of the political group of that city, how the internal disintegration was clear, also verifiable by the orders that the Party issues to the Federations. And also, Foschini had for certain the news that a large column of English troops was on its way towards Kenya, in order to begin the recovery of the Ethiopian Empire, and that, given the difficulty of transporting troops on our part, this English expedition of which the Negus would also be part, would have an easy advance into the territory to be occupied.

And also, that the Germans are rightly indignant at the fact that their victories are morally degraded by the continuous losses suffered by Italy; and that Italy itself should not have distorted the true situation regarding the scarcity of its armaments.

And 3 months later my grandmother returns to the attack:

Rome, 2 March 1941Mrs. Rosa Caccioppoli said on the afternoon of the 26th that the Duce had left for Albania. And that some generals, friends of hers, had laughed when they learned something of the sort because the Duce deludes himself into thinking that his presence can speed up our offensive; and that the presence of the Duce would have a contrary effect as most of the Army is completely hostile to Mussolini.

I don't know how much my father, who was a fervent fascist, was truly convinced of what the secret police had reported. But I find it a singular coincidence that, after the breakup of all relationships between Giuseppe Caccioppoli and my grandfather Lorenzo, his wife Rosa and uncle Peppino's son, Renato, are all together once more, collected by the diligent officials of the OVRA, in the same folder.

14

Caffè Vacca
1936

Neither Sofia Caccioppoli nor Maria Bakunin like Sara Mancuso. The two sisters turn up their noses. The girl is different from all the young people who grew up in their milieu. She is too made up, too casual for her age, and above all she hails from a distant world. It's not just France, the country where she has lived for a long time, that disturbs the ladies, but also the nearby Vomero. On her return from Nice, Sara's mother, Antonietta Castaldi, opened a guesthouse, Alla Santarella, in via Cimarosa 47, where transient students and employees stay.

Everyone, not just in the modest family hotel, notices Sara. Tall, slender, brunette, blue eyes, she is a radiant beauty, but also elusive, haughty. The words that come out of her mouth are sultrily embellished with a sensually trilled r, which in times of chauvinism arouses some suspicion.

I found only one photograph of Sara in those years, which was generously offered to me by her grandson. The image is worn, stained, wrinkled, as if it had been kept for years in a wallet, as was the custom then. But despite neglect and wear and tear over time, all her attractiveness shines through in this photo (Fig. 14.1).

Renato, who is 34 years old, met her at Caffè Vacca, a bar in the Villa Comunale where many friends, like him, who are intolerant of the regime spend hours in endless conversations. She, who is just 16 years old, is in the company of Arturo Labriola, the former socialist deputy, who sensationally left the anti-fascist movement in 1935 and recently returned from exile in France.

They are a couple that sparks speculation and gossip. The over 60-year-old Labriola and Sara just out of adolescence. But the girl is far less naive than her

L. Foschini, *The Friction of Life*, https://doi.org/10.1007/978-3-031-65262-2_14

Fig. 14.1 Sara Mancuso

Neapolitan peers; she has lived in another country, has studied and knows French literature, loves Rimbaud and Baudelaire above all; she is free, emancipated, non-provincial and surrounded by an exotic aura, scandalous and in some ways elusive. She has all it takes to please Renato. She is light years away from the "living dead" he abhors and comes from an indistinct world very far from the female role models he grew up with: his mother, his aunt, but also the emancipated Maria Del Re and the adorable Nella. The budding bond between the young mathematician and the girl arouses anxiety and restrained agitation in the big house in viale Calascione where Sofia reigns like a Russian grand dame, receiving her friends, her friends' children, Russian refugees and Polish relatives in a "slipshod and grand" way. Sofia has followed her son's hypersensitivity with apprehension since he was a child, his frailties, his increasingly frequent periods of prostration, and tries in every way to protect him, as does her sister, Maria, holder of the chair of Organic Chemistry at the University of Naples, granitic vestal of the memory of her father, Mikhail, who is overflowing with admiration for her nephew. But it's all useless. Sara is special and Renato is won over by her. It's not just mutual attraction, but a common desire, strong and rooted in freedom.

15

Ettore Majorana
1938

It is January 10, 1938, when Ettore Majorana arrives in Naples to take up the chair of Theoretical Physics for exceptional merit. The head of the Physics Institute is Antonio Carrelli. Caccioppoli knows him well, not only for academic reasons, but also because, like him, Carrelli attended the Posillipo club, animated by the marchese Pietro La Via as long as the regime allowed it. The presence of a member of Enrico Fermi's Roman working group arouses great interest in the university.

Majorana, preceded by his fame, gives the inaugural lecture on the 13th which Renato probably attends as a teacher with the faculty of science. The course starts on the 15th. The lectures, followed by very few students, are attended by Caccioppoli's assistant, Don Savino Coronato. The priest was certainly sent ahead by his professor to seek out common ground with his Sicilian colleague. In fact, there were boundaries between the two disciplines difficult to cross without fear of arousing the possible envy and suspicion of colleagues.

Even without yielding to a facile symmetry between these two precocious geniuses, one cannot help but notice how their simultaneous presence in the Neapolitan university was exceptional, albeit for a very short period of time.

There is only a 2-year difference between Renato and Ettore. Like Caccioppoli, Majorana, born in August 1906, also studied engineering before shifting, on the threshold of his final year, to the degree course in Physics under the direction of Enrico Fermi. And it is also said of Ettore that he was *strange*, afflicted by recurring nervous breakdowns, that his appearance was bizarre, and that for 3 years, between 1934 and in 1937, he stayed locked in his house studying, receiving no one and rejecting his correspondence. And

© The Author(s), under exclusive license to Springer Nature Switzerland AG 2024
L. Foschini, *The Friction of Life*, https://doi.org/10.1007/978-3-031-65262-2_15

that, like Renato, he neglected his physical appearance and let his hair grow excessively long for a certain period.

Laura Fermi gives a description reminiscent of those later made of Caccioppoli: "In the morning, while taking the tram to the Institute of Physics, he would begin to think with a frown. If a new idea came to mind, or the solution to a difficult problem, or the explanation of certain experimental results that had seemed incomprehensible, he would rummage in his pockets, take out a pencil and a packet of cigarettes on which he scribbled complicated formulae…" It's the same scene that Neapolitans were to witness over the years, following Caccioppoli's daily route from his home to the University.

It has been suggested that the two met and had brief encounters during the Sicilian physicist's 3-month stay in Naples. Actually, however, we don't have any proof of this and all that remains is regret for a missed opportunity. But Sciascia, once again outlining Majorana's characteristics and genius, writes words so perspicacious that they also apply to the Neapolitan mathematician. The genius "obscurely senses in everything he discovers, in everything he reveals, the approach of death; and that 'the' discovery, the complete revelation that the nature of a mystery assigns to him, will be death. He is as 'one' with nature like a plant, like a bee; but unlike these he has a margin, albeit small, for play; a margin within which to get around and circumvent it, within which to search—even if in vain—for a crossing, a vanishing point".

Majorana finds the "vanishing point" on the evening of March 25th, 1938, when he leaves his room at the Bologna, the Neapolitan hotel where he stays, in via Depretis 72, and embarks aboard a Tirrenia steamship bound for Palermo. He stops for a couple of days at the Grand Hotel Sole, from where he sends some letters to Carrelli and family. The next day he leaves again, perhaps, for Naples. But from this moment on nothing is certain anymore and the Sicilian physicist disappears into thin air. He is 31 years old. Legends sprang up around him, but Majorana was to remain forever shrouded in his secret, impenetrable to this day, "a secret, escape from which would have amounted to an escape from life, an escape of life".

16

How to Dispel a Myth
October 23, 1938

Legends have sprung up around Renato, too. He has been attributed with sensational gestures that he never made. And despite not having physically disappeared, like the scientist from Catania, he is shrouded in an aura of mystery to this day.

Among the secret police papers kept in the State Archives, I search with great curiosity for traces of what has been written several times in books and newspapers: Caccioppoli in the company of the young Sara Mancuso challenges the fascist police by singing the *Marseillaise* on the day of Hitler's arrival in Naples, on May 5, 1938. An episode that would have led to his arrest followed by hospitalization in a psychiatric clinic. How many times I have heard this story! It was even rumoured that it influenced a scene in the film *Casablanca*. He being played by Humphrey Bogart and she, Sara, by Ingrid Bergman. According to these fantastical reports the event, that some think took place at the Grottino di Mergellina, and others think in the Löwenbräu brewery in Piazza Municipio, marked the birth of an authentic rebellion.

A group of militiamen asks the orchestra to strike up the fascist anthem, *Giovinezza*. In the room still full despite the late hour, the performance of the song inflames souls already excited by the imminent arrival of the Führer. Renato, intolerant of all this enthusiasm, suddenly gets up and, approaching the piano, asks if he may play it. But, with that masterful touch of his, instead of performing his beloved Debussy he starts singing the notes of the *Marseillaise* with Sara who, having come next to him, begins to sing it in her perfect French. Having finished singing the anthem, Caccioppoli begins to speak, explaining to the astonished onlookers what it means to live in a free country, directly calling into question "*cavaliere* Benito Mussolini, who instead of

L. Foschini, *The Friction of Life*, https://doi.org/10.1007/978-3-031-65262-2_16

being a teacher has established a despotic and illiberal regime." Only to drastically conclude: "It's time to put an end to these barbaric and shameless fascists!" A moment later the militiamen realize the enormity of the affront, they jump on him, stopping his mouth with a handkerchief. Lifted bodily (not difficult given his excessive thinness), Renato is taken away together with Sara. The police station is a stone's throw from Piazza Municipio, but during the short distance the fascists threaten to give him a kicking if he doesn't stop talking. Caccioppoli replies with his mordant logic: "I get it, kicking is the weapon of donkeys!" He is thrown into a holding cell together with Sara and locked up for months in a psychiatric clinic.

This is the story that has been credited for years and that I try to find evidence for among the secret police papers, but in vain. The episode was completely fabricated, and it would be interesting to understand who created this fake news and why. Why has Caccioppoli become a legend, over and above the great merits of his mathematical, musical and human genius, to the point of attributing him with exaggerated and slightly quixotic words and gestures that he had never made and were absolutely foreign to his nature, his sense of the ridiculous, and to his bitter awareness of the tragedy that Italy was going through?

The events—I discovered—took place in months and days distant from the date of Hitler's arrival and went very differently. I have three reports in front of me. The first consists of a very detailed note from the prefect of Naples Giovanni Battista Marziali addressed to the General and Reserved Affairs division of the Ministry of the Interior. The second is a "Note for H.E. the Minister from the Director General", while the third is a report of the events addressed to the Rector of the University of Naples, the senator and professor Giunio Salvi, from the provincial party secretary, Eduardo Saraceno.

The episode that actually happened therefore dates back to October 1938, 5 months after the Führer's visit to Naples. Here are how things actually went.

It is October 23, 1938. In a modest trattoria, Il Grottino, at number 170 on the Chiaia Riviera, at nine in the evening a couple enters and attracts the attention of the customers. The woman is "a very young and elegant lady", while her companion appears to everyone to be "shabbily dressed".

It's Sara and Renato. The two occupy a table and begin to drink two, three, four glasses of wine; then suddenly she, evidently already tipsy and raising her voice to be heard, invites those present to join them: "I'll buy everyone a drink," she says out loud. "I'm happy!". The proposal is accepted with enthusiasm and the few customers, simple people, workers, artisans and unemployed, sit at her table.

Renato says nothing, Sara, on the other hand, talks continuously. She says, with a pronounced foreign accent, that she is French and that her companion is Russian. And after yet another glass, with exaggerated enthusiasm certainly due to too much alcohol, she turns to the astonished onlookers by declaring emphatically: "You are all my brothers!" Music is coming from the radio. The girl grabs the arm of the owner of the club and drags him onto the dance floor. The other customers gradually join them, except for Renato who sits drinking.

It's getting late and it's almost closing time when in comes a certain Federico Manna, tobacconist, in the company of Gaetano Catalano, barber. More and more excited, Sara notices the fascist badge they both wear in their button-holes and, pretending not to know what they are, asks for explanations. The girl's question and general behaviour are clearly provocative and her free and unconventional attitude arouses the owner of the club's suspicions, and he calls the *fiduciario* of the local fascist group.

It's half past ten by the time the couple heads for the exit. The tobacconist and the barber, prompted by curiosity, offer to accompany them for a stretch of the road. They are joined by Salvatore Milucci, street porter, to whom the young lady, in the name of brotherhood, offers to find a job, writing his name in a notebook. The unusual group heads towards the funicular that leads to Vomero, a short distance away.

Once they arrive, Sara says goodbye to Renato, who heads home, and sits in a compartment with the barber, the porter and the tobacconist, who tells her that he removed the badge from the collar of his jacket and revealed that he was against the regime. Encouraged by this statement, the girl falls into the trap and out loud, visibly moved, she declares she feels close to the poor work-ers: "I am sorry for you, you are ignorant and oppressed," she allegedly said. "The newspapers do not tell the truth. It's impossible to get along with fas-cists!" Not content with this escapade, when Sara arrives at Vomero she invites her companions to join her for coffee in a bar. And there, too, according to the witnesses, she continues in the same vein.

However, not everyone agrees with this version. Some, as reported in the police report, defend her: "Catalano, when questioned, declared that he could not say one way or the other regarding the conversation held by the young lady with Manna, as it had taken place between the two in a low voice. He added however, when the young lady took her leave from him, she had said sarcastically: '*Arrivederci*, great Italian and great fascist.' [...] On being ques-tioned, Milucci who, as mentioned above, had left the tavern together with the young lady, declared that he had not heard her say anything of a political nature. [...] Another four customers with whom the young lady had become familiar were interviewed and everyone unanimously declared that the young lady appeared friendly to everyone and that she had not made any judgement, even when she had asked some of them to explain the badges they wore in their buttonholes.

[…] The young lady was identified as Mancuso Sara, an Italian citizen, who had settled here 2 years past and had lived from an early age in France, as her late father had been employed as a hotel manager in Nice. […] The aforementioned has no criminal record whatsoever in the files of the local Police Headquarters. […] The individual who accompanied Mancuso was identified as Prof. Caccioppoli Renato, whose personal details I state in the subject line, who although registered with the P.N.F. since 1933 holds anti-fascist ideas and is therefore monitored. […] In this regard—as is written in prefect Marziali's report—I refer to the Ministerial edicts no. 500/26675 of 18.9.1933 and no. 500/9301 of 18.3.1937, in which Caccioppoli is stated to be an anti-fascist element and propagator of information hostile to the Regime."

Renato, who is returning home, is quickly traced, taken to the police station and interrogated. The police know him, have been watching him for some time, and are aware of his aversion to fascism. He and Sara are arrested. It is suspected that the two entered the tavern on purpose to get close to workers and raise awareness of their political struggle.

But then things seem to get better: "After hearing the witnesses," it says in the police report, "and considering all the circumstances, this Office has come to the conclusion that neither Mancuso's nor Caccioppoli's behaviour in the tavern had any real political purpose, but was determined by the young girl's frivolous and bizarre character and by Caccioppoli's strange and abnormal nature. And in fact Caccioppoli, apart from his undisputed scientific worth, is excessively addicted to alcohol in private life, shows himself to be an abnormal individual lacking all social eventuality (sic), so having given in the course of the interrogation evident signs of mental imbalance, this Office decided to subject him to health checks, following which, recognized as insane, he was interned in the local Provincial Psychological Hospital. Given this and taking into account that no evidence was collected to corroborate Manna's statement regarding the words that Mancuso allegedly uttered during the funicular journey, the same Mancuso was released and handed over to her mother, subject to a formal injunction to supervise her daughter's behaviour and see to her education in order to avoid any repetition of unpleasant events of this kind. As for Mancuso Sara, who was also cautioned, arrangements have been made for suitable supervision."

These are the events that occurred on the evening of October 23, 1938. Inserted later at different times and places, outside of any real context, and finally transformed into legend. The truth is much simpler, although equally dramatic. Sara, with a marked tendency towards the exhibitionism and reck-lessness of a little girl, is the undisputed protagonist of the story. According to witnesses, Renato does not speak. But without a doubt his silence does not discourage the girl, on the contrary.

Having drunk a great deal, Caccioppoli allegedly went off the deep end during the interrogation. Extremely sensitive, psychologically fragile, averse to any form of intimidation, in the face of requests for clarification made in a crude and brutal way by the police, Renato responded with insolence and furious sarcasm. And under the pressure of questions, he allegedly lost control altogether. This was to determine the decision to admit him that very night to a psychiatric hospital.

Another document, this time signed by police chief Giuseppe Stracca, says: "Having given during the interrogation manifest signs of mental imbalance Caccioppoli was subjected to a medical examination and, recognized as insane, he was admitted to the local mental hospital. While Mancuso, whose strange behaviour has been noted for some time, is still detained for ongoing investigations."

17

From Capodimonte to Capodichino
24 October–30 November 1938

On the hills overlooking the Gulf of Naples, there are still places where Renato lived. Villa Caccioppoli in Capodimonte has the melancholy and desolate air of abandoned homes, which however hold a dark enchantment in the few traces of former beauty. The peeling plaster, eaten away by the damp, reveals the tuff stones of the wall covered in ivy. Bushes and weeds grow all around and hinder access. I look at an old photograph. The house appears to me in its ancient splendour, as when it was when my grandfather Lorenzo lived there while very young, surrounded by a garden full of flowers over which tower the crowns of palm trees. The columns of the portico frame the entrance. Upstairs, from a large terrace overlooking the sea, a woman is looking out, perhaps Sofia, perhaps Sara. Who knows?

Not far from here, still in Capodimonte, on the small hill of Miradois, there is the observatory of San Gaudioso, the grandiose observatory wanted by Giuseppe Bonaparte. It was frequented by Lorenzo's father, my great-grandfather Francesco Caccioppoli, when in the second half of the nineteenth century some very special persons worked there. Such as Annibale de Gasparis, who discovered the 11th asteroid and named it Parthenope in honour of the city. In those years, the director of the Capodimonte observatory was the mathematician Ernesto Capocci, who beat Jules Verne to the punch by telling the story in 1857 of *the first trip to the Moon made by a woman in the year of grace 2057*. They were brilliant, eccentric and rebellious mathematicians, astronomers and scientists who, observing the sky, lived in contemplation of the universe as the most extreme and highest configuration of liberty. And while they studied the stars at night, by day they were making a revolution.

L. Foschini, *The Friction of Life*, https://doi.org/10.1007/978-3-031-65262-2_17

49

Following this triangle of family memories, I arrive on another hill over-looking Naples. In Capodichino there is the Leonardo Bianchi psychiatric hospital, formerly Villa Fleurent, the nursing home where the sculptor Vincenzo Gemito was confined and from which he fled in August 1887. In the long-abandoned mental hospital, in search of something that speaks to me of Renato's passage, I find only desolation, abandonment, and neglect. I walk through the vast empty halls, the long aisles where the iron cots are still lined up with their bedheads now devoured by rust, the windows barred to prevent acts of desperation. It's a place where, many years after its closure, the walls still hold traces of anguished affliction in the writings scratched on them.

Caccioppoli arrives there at dawn on 24 October 1938, after a night at the police station. It is not difficult to imagine how this disturbed the highly sensitive professor torn from his world, enclosed between Viale Calascione and the university, and thrown among the insane who, if they are not tied up, wander around these sad corridors in a daze.

Soon, however, the violent impact is cushioned by the arrival of a piano for him. It is the same one that I saw, standing against a wall in Maria Del Re's living room. Renato plays there for hours, improvises concerts, entertains the inmates who, in the sad rooms of the asylum, have never heard the notes of Chopin's Nocturnes, of Debussy's Mer, or of Beethoven's Moonlight Sonata ring out. But it is not only music that gives Caccioppoli relief. He works for hours stooped over the table in his room working on the problem of the existence of closed convex surfaces (the so-called "ovaloids"). Carlo Miranda goes to visit him to study together. He would remember many years later: "We often met in a nursing home in which he was hospitalized."

There is no doubt that Renato enjoys preferential treatment. Not only can he play the piano and receive visitors, but he can even leave the hospital. Gianfranco Cimmino comes to pick him up every day and they go in his car on trips to the Sorrento peninsula, on the coast, travelling as far as Paestum to relive the splendour of the temples. On the way back, they stop in Pontecagnano at a pizzeria where, with unusual appetite, Renato eats a margherita. Cimmino has the clear impression that his friend has calmly accepted cohabitation with the insane, considering it a curious life experience, and in the passionate conversations he has with Cimmino he interweaves, as always, long musical digressions and complex scientific disquisitions.

The young mathematician is fascinated by Caccioppoli's ability to identify the essential in the study of all the issues that concern the correspondences between functional spaces, going directly to the root of what appears concealed by superstructures fraught with difficulties. Carlo Miranda would describe years later the prodigiously unusual way he arrived at the solution: "He didn't like the work of polishing and refining, but preferred to constantly tackle new problems, and with the brilliant intuition with which he was endowed he was often ahead of his time, opening new paths in scientific progress."

To shed light on the days spent in the mental hospital, his medical records recently appeared in the hospital archives, thanks to a tenacious researcher, Anna Sicolo. The diagnosis made by the psychiatrist who receives Renato upon his arrival and treats him during his stay in the hospital is surprising right from the first words with which the patient is described: "Supranormal intelligence." The rest of the notes are no less interesting. The doctor informs us that "since childhood the patient has presented neuropathic features with a tendency towards eccentricity, melancholy, and contradiction supported in the interlocutory stage. He is tormented by extenuating insomnia and is in an authentic depressive state from which he tries to escape by drinking wines and liqueurs [...] he spends the night in cafes and taverns stunning himself, behaving oddly and giving displays of eccentricity." And not only this. Caccioppoli "has a tendency to astonish with strange and impertinent attitudes [...] He delights in surprising." And in support of what has been stated, the episode I previously mentioned crops up again: "Once he appeared in the public street with a cockerel."

After 2 weeks in hospital, Renato shows signs of improvement. The doctor writes that "the patient is fairly lucid and quite serene, he sleeps sufficiently without medicines but," he adds, "sometimes his behaviour is childish".

And it makes you smile on rediscovering his childish side that we already came across in the delightful letters from Padua to Nella and Lina and which we will find again, surprisingly, even in the darkest years. You cannot fail to be curious and to try to have fun imagining what diabolical pranks he must have concocted at the expense of the doctors and nurses of the hospital.

On 30 November 1938, declared cured of "alcohol poisoning" he was entrusted to the care of his brother Ugo and his family so that his impulse to abuse alcohol is kept under control. After a month of hospitalization, Renato returns home. And my pilgrimage through the hills of Naples comes to an end right on the threshold of his home.

Climbing the hill of Pizzofalcone I arrive in via Monte di Dio, full of sumptuous aristocratic palazzi that conceal very shady gardens, a privilege, as Elena Croce noted, enjoyed by old Neapolitans who had no love for holidays. At the end of the road, on the left, there stands Palazzo Serra di Cassano, a grandiose residence from which Duke Luigi's son, Gennaro, left on 20 August 1799 to go and die beheaded in the market square. At just 27 years old, he was another dreamer avid for freedom who, in pursuing the mirage of the Neapolitan republic, met his death. With him many young intellectuals, bourgeois, and commoners perished, leaving a seed that every so often mysteriously buds in the soul of some inhabitants of this city, but equally mysteriously dies before bearing fruit.

Mikhail Bakunin, when he arrived there in 1865, almost seventy years later, believed he could rekindle that desire in the people, that subterranean angst he considered to be a fire that still burns beneath the ashes of the defeated *napolitana* revolution. But once again all illusions dissolved.

Over the centuries the Neapolitans have risen against the Spanish, the Austrians, the Bourbons. They even fought for five heroic days in 1944 to drive the Germans out of the city. But then?

I think of when Herman Melville arrived in Naples in 1857; he managed to capture the spirit of the city in a short poem that remained unpublished for a long time: *Naples in the Time of King Bomba*. The author of Moby Dick, wandering fascinated among the crowds full of life he passed in the streets, suddenly made a discovery that turned the scenario of cheerful folklore that had enchanted him until then on its head: the cannons of Castel Nuovo were not pointed, as would seem natural, towards the sea in case potential invaders were spotted, but rather at the city, on its inhabitants. Whose collective joy, Melville discovers, conceals "Or even in some a patched despair / Bravery in tatters debonair / True devil-may-care dilapidation?"

Before you reach the pink mass of the old Nunziatella military school, viale Calascione is on the right. On paying two cents, in the years when Renato lived there, you could walk and, after going down the hill, emerge at Santa Maria a Cappella Vecchia. A few steps and you find yourself in Piazza dei Martiri, the elegant heart of the city.

The refined pianist Sigismondo Thalberg lived in Viale Calascione in the second half of the nineteenth century and from the first decades of the twentieth century the well-known antiquarian bookseller, Gaspare Casella, also lived there. His house is the landing point for many writers and men of letters who arrived in Naples, from Anatole France to Curzio Malaparte, from Dino Buzzati to Giuseppe Ungaretti. My grandmother's cousins, the numerous Starace family, live at number 7, and the Caccioppolis at number 16. Opposite, in a building gifted by a Starace to his daughter Rosa, a nun, so that she could make a school, there stands the Nazareth Institute. In the thirties, at one o'clock, when the bell rang the street was invaded by swarms of cheering little girls chasing each other and shouting. Among them there was Isabella Ducrot who now, a fascinating 90-year-old artist, remembers Renato swaying as he walked homeward looking straight ahead, mumbling incomprehensible words, careless of the noisy schoolgirls attracted but also intimidated by him.

And it is to this highly particular street that Caccioppoli returns after 37 days spent among the lunatics.

18

Renato Gets Married
29 June 1939

A note from the ministry informs the rector that "following the incident Caccioppoli was seen again in the company of Mancuso Sara, which leads to the belief that his return to service would not fail to result in problems. It is worth remembering that Mancuso lived for many years in France and is believed to be the lover of the well-known former deputy Labriola."

On December 30, 1938, Professor Giuseppe Caccioppoli forwarded a request to the ministry requesting that his son be granted one month of extraordinary leave for health reasons and 6 months' leave of absence for the same reasons. Subsequently Renato was to ask for an extension of another 6 months, which was granted from 1 June to 30 November 1939. In this period, after leaving the hospital, he finds Sara again and after 7 months he marries her at the Vomero town hall, on the 29th of June 1939.

The doubts expressed by his mother and his aunt were to no avail, as were his father's silences, laden with concern. The family is displeased. Despite their allegedly anarchist forebear, Sofia and her sister Maria are well established members of the best bourgeois society in the city. Giuseppe Caccioppoli was Queen Margherita's private doctor and considered one of the most renowned surgeons in Italy. They certainly came to know about the rumours circulating about the girl and perhaps they were also informed of what is said in the police reports: "Mancuso was found to be a young person of a frivolous and childish nature mainly due to the mental habit acquired in the milieus in which she lived abroad and the lack of restraint on the part of the mother, who is not morally in order either."

And to aggravate this judgement there is the even more insinuating and poisonous note drawn up by the provincial party secretary of Naples, Eduardo

Saraceno, addressed to the Rector of the university, Professor Giunio Salvi, in which—summarizing the event—Renato is involved far more than he appears in the police report. Caccioppoli is defined as a well-known anti-fascist and as such is monitored by the police, but above all he is found guilty of associating with Sara Mancuso, a character of questionable morals, "aged 18, who for some time has been noticed for her strange behaviour".

However, none of this matters to Renato. On the contrary. Sara is beautiful, elusive, nonconformist. And in a world where everyone is regimented, on file, controlled by the regime, she is absolutely free. She is the woman for him.

Once married they go to live in Palazzo Cellammare, in via Chiaia 139. They will sleep in separate rooms. Caccioppoli studies at night and doesn't want to disturb his young wife's sleep.

19

Ovaloids of Assigned Metric
May 1939

Only after a few months was Renato to return to university. This is the place that, together with his new home in Palazzo Cellammare, represents the shell in which, protected from the outside world, he closes himself.

Since 1938, racial laws have come into force in Italy. Before the Grottino misadventure, Caccioppoli apparently tries with some friends to help a Jewish girl. A painter friend manages to have the young girl come from Germany to escape Nazi persecution, with the excuse of acting as his secretary. When the repatriation order arrives, Caccioppoli, together with Bianca and Mimisa Vico, Pietro La Via and other friends, raise money to bribe the Italian policemen who were supposed to keep watch at the railway station in Rome to ensure her departure. The plan doesn't work and, at the Brenner pass, the girl commits suicide by throwing himself out of the window. This tragedy deeply wounds him.

Now, a year later, Renato watches indignant and powerless as eminent professors are expelled from universities. Tullio Levi-Civita, the great tenured mathematician of Higher analysis and then of Mechanics, is removed from his chair at Rome's Sapienza university because Jewish. Dismissed from his teaching post, isolated from the academic world, Pius XI has him appointed member of the prestigious Pontifical Academy of Sciences.

It is for this reason that Caccioppoli decides not to send his work, to which he dedicated himself when he was locked up in a mental hospital, to the Accademia dei Lincei, of which he is a member, but rather to the Pontifical Academy of Sciences and in particular to Levi-Civita:

Naples, 10 May 1939

© The Author(s), under exclusive license to Springer Nature Switzerland AG 2024
L. Foschini, *The Friction of Life*, https://doi.org/10.1007/978-3-031-65262-2_19

Illustrious Maestro, the memoir that I make so bold as to send in for examination contains a solution—the first definitive one in my opinion, as I point out in the preface—to the well-known problem of the existence of a surface closed with ds^2 to an arbitrarily given positive curvature.

I would like this work of mine to be published in the Proceedings of the Pontifical Academy and that you—if you deem it worthy—would grant me the honour of presenting it.

In the meantime, please accept the expression of my profound, unchanged devotion.

—RENATO CACCIOPPOLI

Ovaloidi di metrica assegnata appears in 1940 in *Commentations Pontificiae Accad.*

In these tormented years, Renato never stopped doing mathematics. Those able to read his scientific writings are struck not only by the brilliance of his insights, but by the extreme conciseness of his style, by his ability to condense an unexpected wealth of ideas into a very short space, always grasping the truly essential element, highlighting the origin of things, which is often, as the mathematician Lucio Lombardo Radice explains, "hidden from our eyes by bulky superstructures bristling with arduous difficulties and which instead seems to reveal itself effortlessly and with the charm of a surprising simplicity".

Between 1930 and 1939 Caccioppoli attained numerous, extremely important results concerning the theory of partial differential equations. All contained in a dozen brief publications. These were later to inspire the work of other illustrious mathematicians.

Eduardo Caianiello, who in recent years was to attend his lessons and later was to be one of his assistants writes about him and his prodigious work: "Staring intently or writing a few formulae on a blackboard are not the portrait of a mathematician. The inner toil that accompanies a true creative act is torment, a profound crisis [...] what comes afterwards presented as inexorable logical reasoning is in reality almost always the extreme conclusion of a wander, as of one sleepwalking or possessed, through a hostile labyrinth: the light, if it appears, is a sudden flash, intuition. These are the moments I remember about him above all: when he was, simply, 'absent' in this world."

20

Mikhail's Papers Set Ablaze
September 12, 1943

"Give Vincenzino a kicking because he didn't trim his sideburns." Rough, hasty, surprising words, jotted down in the top right corner of an important scientific publication by Renato's fearsome and eccentric aunt, Maria Bakunin, professor of organic chemistry at the University of Naples. Vincenzo is the janitor of the institute, subjected like many others to the violent and irascible character of the professor, who my grandfather met in Belgium when he was studying engineering in Liège. She arrived there in 1913, sent by the minister Francesco Saverio Nitti to find out about "the direction given to the teaching of Chemistry and the results achieved in terms of teaching and the implications for industry". Maria Bakunin visits eleven professional institutes and it is during one of these meetings that Lorenzo Caccioppoli encounters her. For my grandfather, laconic and ferocious, she was the Witch from that moment on. For all the academic world, who reveres and fears her, she is instead the Lady.

A window marked with the house number 10 will catch the eye of anyone who walks along via Mezzocannone and skirts the side walls of the university. In its place, in the past, there was a small door that led, and this is completely unique, to a private home located inside the magnificent university building. Maria Bakunin lived here for many years, first with her teacher and husband Agostino Oglialoro Todaro, then, widowed, with her niece Giovanna, her brother Carlo's daughter, and finally alone.

Brilliant and original just like Renato, Maria is the first woman to graduate in chemistry in Italy. She marries Oglialoro, 26 years her senior, and after her husband's death she becomes the lover of the brilliant student Francesco Giordani. She is 48 and he is 25. The reports of the police who investigated the links between Giordani and fascism after the war say that the two had (also) married in secret. In addition to being extraordinarily intelligent, Maria has a steely character. "Those who lack the courage to do the impossible, will not achieve the possible," she likes to say, quoting the founder of the anarchist movement.

Every day the professor leaves home and heads to the nearby number 4. In the very short stretch of road along which there appears to be a general stampede of students terrified at the mere idea of bumping into her. She rules the institute of chemistry with tyrannical severity. She works for hours and hours on experiments with very dangerous substances such as nitro-glycerine, which can explode at any moment through simple percussion.

And in fact, every now and then detonations would suddenly come from her laboratory. Anyone on the street at that moment got the impression that a bomb had exploded. Old Mikhail, who loved explosives, would have been proud of his youngest "daughter".

The Lady refuses all compromise. She is intransigent, inflexible. Like her nephew, she too exudes an exotic and legendary aura redolent of the Siberian steppes where she was born, in Krasnoyarsk, in February 1873. Considered by all to be the third child of the legendary founder of the revolutionary movement, she became his proud vestal, the guardian of his credo, the form of which he found in Naples and the term to encapsulate it in Ischia. It was on that island, in Casamicciola, where in a letter to the comrades of 1866 Mikhail used the word "anarchy" for the first time. Upon the death of their mother Antonia, after bitter disputes with Bakunin's followers, Sofia and Maria become custodians of the paternal documents, which since that day were kept in a large bookcase in the apartment in via Mezzocannone.

On 12 September 1943 German troops force an entry into the professor's house and set fire to her bookcase. To save her father's papers, Maria braves the flames and refuses to leave the place, now half-destroyed, until the Germans, in the face of such temerity, abandoned the field. The damage is extensive, the losses irreparable.

Renato's mother writes to Benedetto Croce's wife on 25 October 1943:

My dear Adele,

You will have heard about my sister's tragedy. On 12 September, following the killing of two German soldiers in piazza della Borsa, in retaliation before the terrified eyes of the poor woman, the Huns scientifically and systematically burned her entire house, preventing the victim from taking anything away from the flames with their machine guns pointed at her to stop her from saving anything. My sister lost everything she owned, everything she had accumulated with love over a long working life: useful things, beautiful and precious things, which were particularly dear to her and which cannot be replaced and, what astonished us most, our father's documents, delivered by the party to his widow, containing autograph manuscripts by the most important personalities of the time. My sister is as if stunned [...] I immediately had her rent a furnished apartment in via Mezzocannone, but Marussia refuses to go there and says that it is less painful for her to live in a room of the Chemistry Institute adjacent to her destroyed house, a bare and inhospitable room from which you can contemplate your old home and live "in the memory of it" [...] I am telling you this sad story because talking about it gives me some relief and because I feel you understand me. I also consider that in such a sad circumstance there was an absence of that tribute of sympathy, of interest, of homage, which my sister deserved. The Rector did not write to her nor go to pay her a visit while the particulars of the case were not reported to the press, a new document of the atrocities of the enemies of humankind. If the senator were to write about this to Prof. Omodeo the measures thus far omitted could be taken now. I entrust this task to you, my dear and good friend. I embrace you and entreat to love me as much as I love you. Give my respects to the Senator who guides Italy today.

Your most affectionate

—SOFIA CACCIOPPOLI

Benedetto Croce steps in and there is no lack of solidarity and participation.

Maria is now 70 years old. She has been teaching since she was 20. In her view, her profession, her family, her nephew Renato who she dotes on, all things are linked to the dark, grandiose university building. Her universe is enclosed between those wide, grey walls. Always fighting, tireless. In particular against anyone throwing shade on her illustrious parent, making insinuations about his impotence and the dubious paternity of his children. The most sensational case exploded in the 1930s but continued for over 20 years, and Marussia was to dedicate not only all her energies, but also a large amount of money in a strenuous effort to silence it.

Hélène Iswolsky published a biography of Bakunin for Gallimard in 1930 in which she talks of his unconsummated nuptials, his wife's lovers, and of Gambuzzi as the true father of the three children. Maria gives herself no peace. She leaves for Paris. She gets Gallimard to censor those sentences deemed offensive on unsold runs, but the matter turns out to be long and complicated. The battle lasted until the 1950s, when copies of the book arrive in some bookstores in Naples. Maria and Sofia ask Renato and his friend Giorgio Amendola to buy them all, so they would disappear from the city. Caccioppoli indulges her, while commenting in his own way with scathing witticisms that arouse the embarrassment but also the hilarity of the communist leader.

Carmine Colella, who dedicated a large and well-curated publication to the professor, says: "Did Maria know, pretending not to know, or were hers only doubts she did not want to give credit to?"

It is probable that both she and all the other members of the family had suspicions but tended to dismiss them. The call of the charm of the Russian anarchist was too strong for them and the city as a whole. The illusion is too seductive that the boiling blood of the revolutionary might flow through their veins. "I have no certainties, at most probabilities," Renato likes to say, perhaps in light of family "things". And it is possible that this doubt crept into his fragile self-awareness, further contributing to undermining all remote certainty, increasing that indistinct torment, a covert and continuous worry that since his childhood has obscured a mind illuminated only by mathematical insights and the joyous shock of musical vibrations. " I am an erroneous derivative," he appears to have repeatedly told his mother Sofia and his aunt Maria, who knew no peace. For years they have watched over this very intelligent teenager, intolerant of the banality of the world around him to the point that he wishes he didn't want to be part of it anymore, to the point of wanting to die.

Just as Walter Benjamin found in Proust, I find an unexpected aspect of Caccioppoli's education. Renato too, despite his excesses and apparent disenchantment, is like Marcel, a *fils de famille*, attached to his mother, to his fearsome aunt Maria, to his brother Ugo and, viscerally, to his city, Naples, as Paris was for Proust. A geographical and sentimental perimeter where his neurosis finds space and breath.

21

Mario Alicata
1944

At the end of 1944, a lanky, extremely thin man, already balding despite his being only 26, arrives in the city. He always wears a big black hat with wide brims, and an overcoat that is long and dark, slightly tattered but elegant. The young Neapolitan communists and the few staff of the newspaper he is preparing to edit, *La Voce,* are impressed, but above all intimidated. To Naples, recently awakened from the provincial torpor in which fascism has enveloped it, Mario Alicata brings the echo of the political and cultural events that are currently going on in Rome. His fame precedes him.

He fought in the armed struggle against the Germans, he had gone underground, he had been in prison after his arrest in a theatre in Cinecittà, a sign that he is not just a comrade but also a refined intellectual. Friend of Carlo Muscetta and Natalino Sapegno, he is among the screenwriters of one of the films that marked the year preceding the birth of neorealism, *Ossessione* by Luchino Visconti. In Naples no one has seen the film yet because the censors have confiscated it, but there is a lot of talk about it; it is said that it is a masterpiece and that it tackles themes never dealt with in deco films: adultery and even homosexuality. In this regard it was still not known that, during filming, Alicata had written to Giuseppe De Santis, the other screenwriter, enjoining him to "keep an eye" on Visconti, insofar as he had got the impression that the director was moving away from the political intentions that were, at the beginning, the basis of the film. And with Elio Vittorini, in those same years, in an article in *Rinascita* he professes a conviction that art must help "men in a consequent struggle for justice and freedom".

This says a lot about the character and temperament of the communist exponent who, more than a simple party official, is already a leader in his own way. In fact, shortly thereafter his *cursus honorum* begins, which, starting from the municipal council of Naples, was to take him over the years first to Parliament, then to the editorship of *L'Unità* and, finally, to be part of the secretariat of the PCI.

Alicata is not likeable. He's passionate, he's brilliant, but he's not likeable. He dominates the editorial staff of the newspaper, giving the impression of knowing all there is to know. He tackles problems head on and for him everything is either black or white. He imposes his ideas in a manner midway between the jocular and the serious, but then he comes out with violent outbursts, And it is difficult to elude his ever definitive decisions that he argues in a strange accent, a mix of Sicilian, Roman and Neapolitan. In the early days he confides in no one and has lunch on his own in an institute for the deaf and dumb near Palazzo Cellammare, where, sitting at a long table, facing the wall, people eat only beans. But after a few months the newspaper he edits enjoys an extraordinary success. Alicata begins to open up, to go out with his comrades when not at work.

When did he meet Sara? It's difficult to say, even though the milieu they both frequent is the same one, the small local nucleus of communist intellectuals. Perhaps at the home of Paolo Ricci, chief reporter of *La Voce,* who initially hosts Mario in his study in the splendid Villa Lucia, or perhaps on Capri, which in those years could only be reached on fishing boats.

Caccioppoli and his wife love the island, for the moment preserved from tourism. They stay at Villa Pina, the work of comrade Carlo Talamona, creator of the most fascinating and original houses on Capri. The nuclear physicist Sergio Valente, Talamona's nephew, recalls the professor receiving students who join him from Naples to sit their exams. "When he was in a good mood," he recalls, "he would give them an 18 and send them away." In Capri Renato bought land to build a house for himself and Sara. The plan is to divide the plot in two and sell a part, using the proceeds to build the house the design of which is entrusted to the architect and friend Luigi Cosenza. Talamona takes care of everything.

But Renato was never to live in the Capri house. A few years later, comrade Alicata was to live in the villa with Sara.

22

Salon Caccioppoli
1945

Caccioppoli finds out about the relationship very soon. Sara tells him about it, even though the decision to leave him was to come a few years later.

His wife is with him when Gide passes through Naples in 1945. The writer notes this in his diary. Eight years have gone by since the encounter in Sorrento, but the two have always remained in touch. And they also saw each other again, as Robert Levesque recounts in his *Journal*. When Gide left France in 1939, on his way to Egypt, he stopped over in Naples and met the young mathematician. And now that the war is over, Gide sets out on a journey once more, still with Levesque, heading for Cairo. And he passes through Naples again. It's December. The city strikes him as being at the end of its tether, buried as it is under the rubble. The hotel Patria where he lodges is run-down, inhospitable, cold. He therefore decides to move to the La Sirena hotel, near the central station, requisitioned by the British army. "Out of the frying pan and into the fire," he notes. "We dry ourselves with the bed sheets. I feel dirty, filthy, punished for wanting to change places."

This time too he meets the Neapolitan professor: "Here I met Professor Caccioppoli again. With whom I spent an unforgettable evening in Sorrento in 1937. Dinner—a banquet at his mother's (Bakunin's daughter's) house with his wife and brother." The two are deep in conversation. The writer falls under the charm of Renato, who talks for hours, tirelessly. The war, the fall of fascism, but above all music, which Gide, who once studied and then abandoned it, loves as much as he. "When I think about my farewell to music," he wrote in those days, "I almost feel my heart is missing and it seems to me, now, that death cannot take away anything I cared about more." That evening Renato plays for him.

L. Foschini, *The Friction of Life*, https://doi.org/10.1007/978-3-031-65262-2_22

A few months before, following the liberation, Eduardo De Filippo manages to return to Naples with makeshift means. He finds the city, although plagued by the wounds of war, pervaded by an almost festive atmosphere full of ideas and hope. Friends that he meets appear animated by an enthusiastic desire to get things done, eager to roll up their sleeves in the illusion of participating in the new Italy that is about to come into being. Shortly after his arrival, in a gallery in via dei Mille where young artists exhibit their works, he meets Caccioppoli. Eduardo is homeless, Renato enthusiastically offers him hospitality in his apartment. De Filippo only stays there for a few days, but in recollections of many years later he did not hesitate to define them as wonderful. They talk about everything, especially theatre. Caccioppoli amazes the artist. He knows his plays and discusses them with such competence and depth as to inspire Eduardo to give him a project that he is thinking about. After the initial euphoria that overtook him when he arrived, he tells him that Naples seen up close, on the contrary, appears enveloped in an atmosphere that he does not hesitate to define as "ephemeral abundance and extreme bewilderment", and in the long nights that they pass, sleepless, he explains to him the plot of the play that is taking shape in his mind, *Napoli milionaria*. Every now and again Renato interrupts him and sits at the piano. He plays as if carried away by the fervour aroused in him by the story in which the protagonist, Gennaro Jovine, returns home to find his family in the grip of a desire for money, and unwilling to listen to his war experiences.

After a few days De Filippo leaves the house in Palazzo Cellamare and moves into a small studio apartment that he has rented, but he doesn't stop seeing his new friend. They eat together every day in the trattorias in the centre, always conversing intensely for hours. Eduardo continues to describe to Renato the characters of the Jovine family who are gradually taking shape: Gennaro, Amalia, Amedeo and then Donna Peppenella, Errico *Settebellizze*, so real and touching in their struggle for survival. Figures that embody so profoundly the drama of a city and its inhabitants that Renato and Eduardo know better than anyone else. Beings lost between an ancient innocence and an unstoppable moral decadence. A tragedy that De Filippo summarizes, in *Napoli milionaria*, in the all too famous phrase "*addà passà 'a nuttata*" (the night must pass), but above all in the bitter observation that, for the Neapolitans, "*'a Guerra nun è fernuta*" (the war is never over).

23

Salon Kosagovskaya
1949

The city breathes beneath the blanket of debris from the past. The rubble of the buildings overlooking the port, which collapsed during the bombing, were to show their skeletons for years and years. Some, without the main facade, look like dolls' houses, open at the front to give little girls the opportunity to play by moving their tiny dolls between one room and another. In this way the buildings on Via Marina reveal to passers-by traces of an old domestic intimacy. Rooms with three walls left miraculously standing, peeling or covered with torn wallpaper, dramatically reveal fragments of life interrupted: in one corner an armchair nibbled by mice, on the floor a worn mattress and even a chandelier that hangs over the now long-gone dining table.

But if from the marina, crossing Piazza del Municipio, we proceed from San Ferdinando up via Chiaia, this mysterious and unpredictable city hides unexpected wonders amazingly left intact. The tragedy of war, which left wounds and scars in the ragged and hungry population, seems to have preserved some, unscathed as demigods, immune to every wound. On the outbreak of war, many aristocrats, the princess of Cerenzia, the Duchess of Marsiconovo, but also bourgeois ladies such as my grandmother, closed their doors behind them and left for Sorrento or Ischia as if they were going on holiday. They came back 5 years later and found everything intact, including the porcelain. the Capodimonte biscuit, and the *bois de rose* display cabinets all in place. A Russian duchess of Jewish origin, the wife of an ambassador of the Tsar who fled the revolution of 1917, returned to her beautiful apartment

L. Foschini, *The Friction of Life*, https://doi.org/10.1007/978-3-031-65262-2_23

in Palazzo Cellammare at the end of the war and started receiving guests again as if nothing had happened.

Roberto De Simone, on the threshold of ninety, remembers to this today the house, the guests, and the concerts held there with an amazing freshness of detail. But, owing to the strange circuits of memory, he forgot the name of the fabulous lady. Intrigued, I perhaps managed to trace the identity of the mysterious aristocrat: it was Sofia Josafovna Kosagovskaya. De Simone, the author of *Gatta Cenerentola*, says that in the Russian lady's living room he, then a very young student at the conservatory, met Renato Caccioppoli for the first time. "The link was my teacher of composition, harmony and counterpoint, Renato Parodi. As well as teaching at San Pietro a Majella," De Simone tells Corrado Valletta who interviews him, "maestro Parodi had graduated, moreover with an excellent mark, in pure mathematics and Caccioppoli—I think, newly appointed to the chair—was his teacher." It is in San Pietro a Majella that Parodi mentions Renato's name to his pupil for the first time: "I can teach you counterpoint," he tells him, "but Caccioppoli can teach you better than I." And the musician enchants young Roberto with enthusiastic stories about the mathematician whom he describes as brilliant, fascinating, bizarre, but also severe and fearsome. Thanks to Parodi, the conservatory student is invited to play the piano at the duchess's house.

One evening, as soon as the concert ended, Caccioppoli comes up to him. The boy is intimidated, even afraid, so vivid is the impression left on him by the stories about the professor, but Renato approaches him politely and starts talking to him with extreme kindliness. Caccioppoli observes that he found the Chopin ballad he has just heard most beautiful, and he regrets, despite having studied so much, that he doesn't have the technique to play it. From that moment Roberto and Renato meet often at the noblewoman's house, at San Carlo, but also for long walks.

The professor enjoys surprising the student with salacious stories or takes metal rings out of his pocket to improvise a magic trick. He challenges him to put one inside the other, but only he is capable of it. His conversation is enthralling, it ranges over the most varied and unexpected themes, from Hermes Trismegistus, who according to tradition was responsible for the birth of alchemy, up to the miracle of San Gennaro. DeSimone discovers with surprise that the professor is no stranger to the knowledge of certain rituals of an ecstatic nature, including the Orphic tradition. "Although he approached these topics with great lightness and irony," he says, "it was obvious that he had great expertise and respect for the subject, as I ascertained over time when I became interested in popular religiosity."

After months these meetings end for no apparent reason. "He disappeared," De Simone concludes his memories of him, attributing the departure to the concomitant end of Renato's love story with Sara. "I met him in Via Chiaia, when he was completely drunk. He was wearing a very dirty raincoat and was staggering. I didn't approach him, out of decency: even if he had been lucid enough, I don't think he would have wanted to greet me in that state."

24

Salon d'Avalos
1950

Every day, Renato crosses at his own light pace the network of streets that wind behind Corso Umberto I dominated by the majestic façade of the university. The alleys, noisy and teeming with humanity, are like veins pulsing in the living body of the city: San Biagio dei Librai, San Gregorio Armeno, Spaccanapoli and Piazza Bellini, on the Roman *decumanus maximus*, now Via dei Tribunali, with the remains of the walls of Greek Neapolis which recalls the origin of ancient Parthenope. Just go past the pizzeria on the corner of via Costantinopoli and take via San Pietro a Majella, and after a few steps on the left you can see the grey outline of the conservatory.

It's a surprise to go beyond that gloomy door and discover that there is a world apart, apparently asleep in its intact beauty, but where on the contrary the air resounds with the light and lively confusion made by the instruments of the students intent on doing their exercises. Caccioppoli often goes there. He meets Parodi, maestro Vincenzo Vitale, and the pianist Marta de Conciliis. He attends the concerts held in the beautiful room that years later was to be completely destroyed by a fire.

San Pietro a Majella is the other side of that eternal Piedigrotta that is the working-class neighbourhoods of the city. Since 1826, when Francis I brought together the pre-existing musical institutions, Francesco Cilea and Gaetano Donizetti succeeded one another as directors, lecturers, students and, when he was already old, Giovanni Paisiello. In its cloisters, among the rooms with a wealth of old instruments, you breathe a rarefied atmosphere suspended in time, and going through those miraculously intact places is an experience that goes beyond aesthetic pleasure. Like the observatory of San Gaudioso in Capodimonte, the conservatory embodies the hidden, secret soul of Naples,

© The Author(s), under exclusive license to Springer Nature Switzerland AG 2024
L. Foschini, *The Friction of Life*, https://doi.org/10.1007/978-3-031-65262-2_24

the austere one, cultured to the point of erudition, but also anarchic and rebellious. The echoes of the revolution resonate in the marvellous library, where the anthem that Domenico Cimarosa composed for the Parthenopean republic and for which he was sentenced to death upon the return of the Bourbons. It was Lord Hamilton who saved him.

In the vast rooms decorated with neoclassical stuccoes, flooded with light, Vincenzo Bellini and Saverio Mercadante taught or studied. There, the high notes of the famous castrato Farinelli, pupil of Nicola Porpora, resounded. De Simone studied there and, first as a student and years later as a teacher, those rooms were frequented by a cultured and refined 20-year-old, very musically gifted (he was to become a highly esteemed composer) but filled with an anxiety and a nostalgia for the past that make him a particular and elusive character.

Prince Francesco d'Avalos is the complete opposite of Renato Caccioppoli. While the latter is an unruly genius intolerant of any form of authority, the aristocrat rejects all marks of modernity and takes refuge in the memory of Naples as the capital of a lost kingdom destined never to return. Since he was 12 years old he studied piano first with Marta de Conciliis, then with maestro Vincenzo Vitale, and composition with Renato Parodi who once again is the *trait d'union* between Renato and him. In the hall of the villa of Posillipo, from the enormous windows wide open over the sea, Francesco receives guests on Thursdays and Sundays for musical afternoons. You don't play, you listen. Through his acquaintances, since the arrival of the Americans at the end of the war, the prince has the possibility to receive the finest recordings, the most refined discs featuring the performances of the great conductors of the time. To hear these, many young music lovers climb the hill on foot or take the 140, the trolley bus that leaves from Piazza del Gesù and goes up to Posillipo.

Another brilliant and imaginative 90-year-old, Federico Forquet, tells me about this. Before leaving Naples at just twenty to join Balenciaga in Paris and start his successful career as a great tailor, he also studies at San Pietro a Majella and frequents the Avalos house, like many other passionate conservatory students or simply music lovers. From the composer and critic Guido Pannain to the writer and biographer of the Bourbons of Naples, Harold Acton, from the musician Hans Werner Henze to Livio Patrizi, to Melina Pignatelli and countless others thirsty for musical innovations. Forquet does not remember Renato's presence in Via Posillipo 47. But d'Avalos and Caccioppoli met at the house of the Russian duchess in Cellammare and they met wherever some important concert took place. So, here too.

What do the reactionary prince and the anarchist mathematician talk about? The thread on which their communication runs is music, with its language that overcomes every barrier of ideology and class. In the cultural life of this city that has known moments of greatness but also many periods of decadence, the importance of music is unheard of in all its expressions, from the highest to the most popular. Naples is not just a city worth seeing in its boundless beauty, but it is also a place worth lending an ear to, a place for listening.

25

The Beginning of the End
1946–1949

We are on the eve of the referendum. Italy must choose between monarchy and republic. In Naples the people are for the king.

Caccioppoli, to quench the monarchist fury that is spreading especially in the poorest streets, has posters printed on which, roughly speaking, it is written: "If the monarchy wins the referendum, it is committed to sweep Naples clean of smuggling, prostitution, and drugs." During the night the professor has his students, including the mathematician Guido Stampacchia and other peers in their early twenties, put up the posters, which bear the king's apocryphal signature, in the most wretched and infamous neighbourhoods of the city, inhabited by criminals, thieves and whores. The purpose, imaginative and burlesque, typical of Renato, is to discourage the ne'er-do-wells among the Neapolitan population from harbouring any monarchic ambitions by ventilating the hypothesis that, were the king to win, of the end of their "profitable activities".

A year later, on 5 June 1947, at 11.45 am Giuseppe Caccioppoli dies in the house in viale Calascione 19. He is 94 years old. The member of my family who was referred to, with a note of regret, as "Uncle Peppino" took his leave with his typical discretion and with the respect with which he was always surrounded.

I found, among my grandmother's papers, the card with his commemorative image showing the dates of birth, December 11, 1852, and of death. I look at his photograph. He no longer has the flowing beard he had when my grandfather frequented his house in Capodimonte but has a drooping moustache running across his face. An elegant boater and dark, narrow eyes reminiscent of Renato's; as for the rest he doesn't look like him at all.

L. Foschini, *The Friction of Life*, https://doi.org/10.1007/978-3-031-65262-2_25

"Professor Giuseppe Caccioppoli," says the writing on the image in the courtly style of the time, "Pioneer and Master of the science and art of surgery. His name is linked to the Scuola degli Incurabili. He reconciled with extreme enthusiasm the temerity of science and the charity of human suffering. He was the benefactor and friend of his patients. Nor did he want public recognition and honour, satisfied with the dear domestic silences where his goodness will remain forever. Consolation and regret to his widow and children."

I have in my hands the letter in which Sofia thanks the literary critic and writer Giuseppe Toffanin, who wrote the text:

September 15, 1949
Illustrious friend and professor
I enclose a copy of the announcement you dictated for our dearest departed.
I intend to repeat it on the tombstone of our chapel in the Pianto cemetery.

In the rest of the letter the lady expresses all her gratitude also on behalf to her son Ugo. The firstborn is not mentioned even in passing. It is very likely that Renato, who abhors all forms of rhetoric, has not read these deliberately affected words. And it's hard to say that from that day onward he felt more alone. In the chronicles of friends and acquaintances who frequent his house, the figures of mother Sofia and Aunt Maria are predominant, while those of the father and brother remain in the background. But this means little.

In the seventies, with Renato long dead, my grandfather's brother, Francesco Caccioppoli known as Ciccillo, section president of the Council of State, when passing through Rome begins to assiduously frequent Renato's brother, Ugo, magistrate. They meet at the chess club and often talk about their families. Ciccillo is a great, cultured spirit, with a surreal and paradoxical ironic streak very similar to that of Renato; he plays the violin magnificently and he too reads musical scores as if they were novels. But what all three share is a great modesty of affections, even though one senses the strong bond with the world they come from. The members of this unusual family, almost an ethnicity apart, have another common element. A complexity in relationships with the female universe made up of attraction, involvement, but also of a subtle fear that often leads them to sarcasm and a sort of irresistible infantile regression in order to protect themselves. But those who believe that this is misogyny are mistaken. Women, all cultured, emancipated, and of marked intelligence are among the most important figures in Caccioppoli's life.

With Maria Del Re, Renato even manages to bare his soul, opening up a chink of light on the incurable turmoil of his spirit. "I have listened to the gramophone a lot these last few days," he writes to her, "the beautiful music I have heard has boosted morale a little (only a little!). The professor's presence, her constancy and also the trepidation that transpires in the letters that she writes to him are such that he feels the need to elude for once the usual irony behind which his emotions find shelter, with words that flow sincere and touching: "How good you have been to me, truly in a 'Christian' manner, I think I only realize now. Like those who only see the trees, and not the woods in which they find themselves, I spoke out against your complaints and I didn't know how to recognize your goodness. Forgive me: to appreciate the line of a landscape a certain detachment is required, as between Naples and Saint Agatha, isn't that so?"

But at a certain point the very delicate balance of Caccioppoli's love life suffers a powerful blow. After almost 10 years, Sara leaves him to go to live with Alicata. It's 1949. A few months after his wife's farewell, Renato is in Paris for the World Congress of Peace Partisans. He knows and loves this city well and yet, to those who accompany him, he appears unsteady, hesitant in his step, almost lost. "Renato was as if lost on the streets of Paris. He had to be taken care of. And he was only 45 years old!" recalls Giorgio Napolitano many years later.

Once his marriage is over, Caccioppoli searches for, and rarely finds, a happy relationship with the female world that surrounds him. He has other women, he is loved. But he is still a man irreparably tried, suffering. A suffering that affects his body, leaving indelible traces,

In June of the same year, he holds a conference in Parma. It is interesting even for non-mathematicians to read his conclusions. "I believe," he writes, "that I have brought to these questions a certain healthy realism typical of us Italians. So, in conclusion, not method, but general direction. A point of view if you like; a sceptic might call it a taste, a politician might call it a programme and, why not?, a poet might call it a state of mind, just as Anouilh said that the landscape is a state of mind, so a complex of theories could ultimately be a state of mind."

This is his way of *feeling* mathematics. And to explain this in years in which the purge is extremely harsh, he quotes Jean Anouilh, accused after the war of collaborationism for having expressed in a newspaper, *Je suis partout*, an orientation close to the German occupiers. The writer defended himself thus: "I have never, not even remotely, sympathized with the Nazis and their sad accomplices, but I confess to having a certain compassion for the vanquished and I fear the excesses of the purge."

It is no coincidence that Caccioppoli mentions him. Like his master Picone, in judging men Renato goes beyond ideological borders. He shares the French author's apparently amused scepticism, behind which lies a deep pessimism.

The protagonists of the dramatist's works are often faced with the harsh choice between ease of compromise and an inflexible idealism. But the compromises are rejected and, with them, life. And the inevitable solution, for Anouilh, is death. His Antigone says: "We are of the tribe that asks questions, and we ask them to the bitter end. Until no tiniest chance of hope remains to be strangled by our hands. We are of the tribe that hates your filthy hope, your docile, female hope!"

26

Whispers and Cries
1950

"My morganatic brother, my brother-in-law": thus Caccioppoli, without ever naming him, indicates the man for whom his wife has left him. He does it with the subtle irony always mixed with mockery that is one of his main characteristics.

At first, in the provincial microcosm of the Neapolitan branch of the PCI, the scandal is a bombshell. It is said that a delegation even went to Alicata to express the party's disappointment at having stolen the wife of comrade Caccioppoli, who eats less and less, is of a ghostly thinness, and above all increases the already large quantity of alcohol that he drinks every day, but is certainly not surprised by what happened. Senator Mario Palermo is also instructed to step in and act as a peacemaker to the couple. An undertaking he soon gives up on. But one gets the impression that, despite the generally obscurantist cloud that has weighed on the party in recent years, so well recounted by Ermanno Rea in his *Mistero napoletano*, within the most elite group of intellectuals, artists, cultured and original comrades, the event was experienced with greater nonchalance.

The slightly morbid gossip, the prurient attitude, the greedy curiosity for scandalous details explodes rather in the bourgeois houses on Via dei Mille, and on the Riviera of Chiaia and Posillipo. Duccio Trombadori, son of Antonello, at the time one of the most well-known communist intellectuals, tells me that he went to Capri as a child with his mother, the fascinating Fulvia Trozzi. It's 1950. On the island they are guests of Sara and Mario Alicata and were to meet them again in that same year in their Neapolitan home. Everything—Duccio has a very vivid memory—is experienced by the protagonists of this story with simplicity, ease and non-conformism. "My

© The Author(s), under exclusive license to Springer Nature Switzerland AG 2024
L. Foschini, *The Friction of Life*, https://doi.org/10.1007/978-3-031-65262-2_26

mother," he adds, "also frequented Caccioppoli in that period. A sign that there were no preclusions of any kind."

And in fact in this story, it is difficult, if not impossible, to imagine Renato in the guise of a Pirandellian character exacerbated by anger and jealousy. His wife is a free woman and Caccioppoli is undoubtedly a man with whom coexistence is not easy, Mario is a winning character. Sara's relationship with Alicata was to continue for a decade. She follows him to Rome and is at his side in his rise to the top of the party even though she sees Renato every now and then. They were still to be together. Theirs is a scarred, difficult relationship, but one that was never to be broken.

In the fascinating and complex psychology of this extraordinary man, the loss of his wife represents a deep wound that can never be healed because what binds him to her is not only love, but a tangle of passions all of the same enveloping and irresistible power: desire for freedom, courage, but also the non-conformism that verges on shamelessness and sincerity that amounts to cruelty. In the background, never resolved between them, the issue of Sara's unfulfilled desire for motherhood. When she gets pregnant, Renato apparently convinces her to have an abortion: "I can't bring another idiot into the world," he allegedly said with his apparent cynicism. The loss of the child certainly marks the moment of definitive crisis in a relationship that stood in fragile equilibrium over time, but one unscathed until then by deceptions, escapes and even betrayals.

On Sundays when many young people flock to the cinema club to listen to him, enchanted by Caccioppoli's compelling presentations of the films, his opening remarks are always for Sara, even though few perceive this. His wife is never absent. Even though they have no longer been together for some time, she is punctually present in the room, sitting in the front rows, as recalled by Gerardo Marotta, husband of her younger sister, Mimì Mancuso. Renato greets Sara in his own way, with words the meaning of which only she and a few others can grasp: "And so every year we find ourselves here like old lovers to inaugurate this cinema club."

The few photographs that we have of her show us an elegant young woman in a light Chanel tailleur at the premiere of *Mandragola*, just off a plane or on Capri beside Neruda, always at Mario's side. Tall, with beautiful, tapered legs, her face constantly in a pout (Fig. 26.1). Never smiling. Palazzo Cellammare is now far off and even more so her mother's boarding house alla Santarella.

At the State Archives I find a note from the ministry informing us that Professor Caccioppoli "is no longer entitled to the family supplement as from 26 March 1956, subsequent to legal separation from his wife".

Fig. 26.1 Sara Mancuso and Mario Alicata

27

On His Trail
2020

From the Central station, where I got off coming from Rome, the university of Naples Federico II is a short distance away and I decide to go on foot. As I walk I revisit what I have read on the train about Renato's scientific work. Carlo Sbordone, who founded a school of mathematical analysis in Naples and has continued and renewed the tradition of classical research undertaken by Caccioppoli, clearly explains that it is impossible in a short space to give a complete idea of his work, which involved the principal sectors of mathematical analysis. "Caccioppoli's primary interests," he states, "were directed at functional analysis and the theory of functions of a real variable, with particular regard to integration and set functions, to the quadrature of surfaces, and to linear approximation."

I reflect on how the scientific world's admiration of his work is understandable, but I still can't make sense of the fascination he exerted on people who know nothing about his research.

As I walk I stop every now and then to think, as if pursuing a thread that I still can't catch, and I'm late for my appointment. Picking up my pace, I pass through the majestic facade of the university to ascend via Mezzocannone on my right, and I arrive in front of the window that indicates with the number 10 the house that belonged to Marussia Bakunin. I realize that, so absorbed in my thoughts, I have gone past the entrance to the institute. I go back a few steps and I am at number 8. Behind the monumental entrance of the university, classrooms, halls, grandiose spaces, endless corridors. The Jesuit Collegio Massimo was located here, the former house of Jesus, the so-called House of the Savior complex closely connected to the nineteenth-century building

© The Author(s), under exclusive license to Springer Nature Switzerland AG 2024
L. Foschini, *The Friction of Life*, https://doi.org/10.1007/978-3-031-65262-2_27

where the first university premises were located, abandoned by the Jesuits after their sensational expulsion following the suppression of the order.

Kind guardian of the place, generous with his time, Lucio Carbone, who occupies the chair that belonged to Renato Caccioppoli, is waiting for me. The professor guides me through a large and narrow seventeenth-century room that must have seen better times, but that now, he explains to me, enveloped as it is in a blanket of white dust, is being restored. From the wide windows the light illuminates Formica furniture with piles of papers, packing boxes, shelves crammed with all kinds of things, tables overloaded with books, tangles of electrical wiring, old unused computers, cupboards that, behind glass doors, reveal piles of paper, files, handouts, rubbish bins half bulging with waste and dumped in a corner. But as I proceed I can already glimpse, standing out in the background behind a large plastic container, a magnificent dark stone doorway overlooking a hardwood door framed by a band of red marble veined with white. In the centre, on a black plate, I read in large golden letters: INSTITUTE OF HIGHER ANALYSIS.

Equipped with a noisy bunch of keys, Carbone opens it and introduces me to a world that immediately seems a thousand miles away from what I have seen so far. Silent, ascetic, miraculously intact. Of a beauty made austere by a Counter-Reformation atmosphere that still lingers in rooms with blinding white plaster. Standing against the walls are heavy, dark sacristy furniture, carved wooden boiserie, fluted columns, and solid walnut seats, the same ones on which the Jesuits recited their prayers; showcases which, instead of the pyxes and chalices of the past, contain a myriad of geometrical solids with round, elongated, pyramidal and elliptical shapes which, although dating back to past centuries, already herald Balla's futurism.

To alleviate the gloom of this series of rooms full of desks, wardrobes and inlaid wooden bookcases, wonderful, large, cheerful windows overlook the roofs of old pink tiles behind which sweet, soothing, protective, Vesuvius stands enveloped in a blue cloud. The most beautiful of these is the French window of a balcony overlooking the port and the ships and past them, from the gap beyond the lighthouse, the open sea. Looking out, I see below me the tangle of alleys that I passed by on my way here and which now seem less chaotic and even more harmonious.

I'm in Renato Caccioppoli's room. Behind the walnut desk, square and imposing, is his armchair with a high Louis XIII-style backrest, upholstered in green leather. Beyond, the bookcase in light wood purfled with gold, closed by opaque glass with Art Nouveau decorations, strikes me as too affected in that very austere context. But certainly, Renato stayed there very little. While I pass through these places that appear to be asleep in search of something that might tell me about him, Carbone begins to illustrate the sectors mostly expressed in his teaching: geometric measure theory, the theory of differential equations, fixed point theorems, the inversion principle and its applications, complex analysis…

We enter other rooms. They are the ones where his assistants worked, especially Don Savino Coronato. Tall, abandoned blackboards still show the white wording of old exercises: square roots, unknown quantities, undecipherable equations that a worn blackboard cleaner has not completely erased. The chalk dust has settled on the wooden frames; on everything, on the desk, the objects, the chairs, a light mantle has settled that the sun illuminates, making it appear like a light blanket of golden particles.

It is said that in certain places the presence hovers of those who inhabited them. I do not think so; I rather think that these presences are within us, and we try to give them form and substance by going back over the places where they lived. They can take on a certain consistency once we find ourselves within the walls where they lived, deluding ourselves into thinking that finally our search is over. But it's not that simple.

28

The Exams
1950–1952

Lucio Carbone opens a small door, I slip in behind him and I climb, in the dark, up a short flight of steps. I emerge into a classroom of a shape I have never seen before, narrow and long. It looks like an alley sloping downhill. There are still the old benches where students, teachers, and curious people attended Caccioppoli's lessons and above all the exam ceremony. His lessons last no more than 25 minutes. The reason is that "only a very few could follow them for longer." Sometimes he teaches them in Neapolitan, because "Mathematics," he says, "is poetry."

I look at the photographs from those years. There is a shot taken in this very room. A large group of students, some sitting at the desks, some standing. In front of them, Renato. The usual newspaper sticking out of his pocket, the handsome, emaciated face, the hint of a smile, eyes narrow as slits. Almost leaning on him, small, slim, with a shrewd and intelligent face framed by crew cut white hair, his black cassock rumpled, is his assistant Don Savino Coronato. A patient and constant presence next to Caccioppoli. The priest and the rabid anti-clerical atheist, author of many jocular quips: "And this lot should teach our children mathematics," Renato shakes his head in front of yet another ill- prepared student, "luckily you and I, Don Savino, don't have any!" Then, looking at him with his usual mocking air: "It's true that we don't have any children, Don Savì? Come on, tell us the truth!"

I look at the faces of the boys, their cheerful intelligent expressions. They appear to me to be aware of living a rare privilege, that of being in the presence of a master. There are more than 20 of them, arranged in four rows. Above, alone, the head janitor Gigino Allocca in an impeccable grey suit with a dark tie, with a severe expression that hints at a vague smile beneath his

L. Foschini, *The Friction of Life*, https://doi.org/10.1007/978-3-031-65262-2_28

bushy moustache and metal-rimmed glasses. He is so taken up by his role that I had mistaken him for a professor. "In this institute only three people know mathematics: Gigino Allocca, Carlo Miranda and, modestly, yours truly," jokes Caccioppoli.

The janitor has known the teachers since they were students. Of great authority, he is the sole distributor of toilet paper. None of the boys allow themselves to argue with him. To those waiting to face the terrible exam with Caccioppoli he asks very seriously in Neapolitan dialect: "Have you studied Shwarze's theorem? No? Then there's no point in showing up."

On exam days, at ten in the morning the classroom used for the 2-year engineering course is already crowded beyond belief. They come an hour early to find a seat. There are students of medicine and literature, coming from courses that have nothing to do with mathematical analysis, others who have already sat the exam. The professor arrives in his rumpled raincoat, grubby, yet very elegant. As soon as he enters, a hurricane of applause. He responds distractedly with a wave of his hand.

The exams are public and take place on the blackboard. Some, the best prepared, say they had a clear sensation that, to overcome the "torture", it was necessary to challenge it. For the others instead it is a torment to be endured until the last act in which the execution, already predicted, takes place. Because if it's an off-day Caccioppoli will massacre them. He is capable of failing up to ten students in a row on the terms of the same theorem and on the calculation of its derived function.

"You're in the kitchen, you need to cook up some spaghetti. The pot with water is on the table. The stove is already lit, what is the first operation you perform?"

"I'd put the pot on the stove," the examinee replies, intimidated.

"What if the pot is not on the table, but on the top of the sideboard?"

"It's all the same," replies the student, "I still put the pot on the stove."

"You are not a mathematician. A mathematician puts the pot on the kitchen table and goes back to the previous case," replies Caccioppoli, and exclaims in loud annoyance: "There's nothing to be done, the dead cannot be resurrected!"

He shakes his head, ruminating on the sloth and paltriness of students and colleagues incapable of higher goals, all intent solely on making their way in the world. Disconsolate, as is his habit, he runs a hand through his hair, his left hand reversed, with the palm up and the fingers like the teeth of a comb to push back the usual indomitable forelock. He gets up from the desk and takes two steps nervously into the chamber. Then he sits down again. To a student who during the exam wants to overdo it to get a higher mark, he says annoyed: "You are required to talk some obligatory nonsense but not the optional stuff."

Renato's intolerance runs deep. It is not born only from exasperation towards stupid people, those who "don't get it". But also, and above all from disappointment, from a hope dashed, perhaps cultivated after the fall of fascism, that something might have changed. And from the bitter realization that everything is still as it was before.

After the war the city woke up destroyed, morally too. The particular tendency of the Neapolitan spirit to live on one's wits has developed the vice that is considered a virtue in Naples: cunning. Everyone seeks to become more cunning in an attempt to survive. Renato, on the contrary, in his originality which is judged madness has aspects of purity and rigour now long lost.

Guido Zappa, called in to teach in 1946, says that Caccioppoli's severity in exams, his intolerance that at first glance might have appeared excessive, are instead strongly linked to this type of disappointment.

Renato is examining a student. She is poorly prepared. At the end of the test, with unusual calm, almost with resignation, he says to her: "You expect to get ahead without deserving it and then get a job through recommendations from the Christian Democrats."

Achille Lauro's right-wing monarchists were in command in the city, but the power of Gava's Christian Democrats was equally strong. These were years of bitter social conflict, workers against entrepreneurs, students against teachers. Caccioppoli explains to his colleague that each party, from their own point of view, is right. "But the solution is impossible," he says. "From which," says Zappa, "Caccioppoli deduces that capitalism, due to its own nature, doesn't work."

The faculty is occupied. There is a protest over fees and exam sessions. The teachers are furious. Renato shows solidarity and sends flasks of wine to the youngsters. His familiarity with the students is surprising. It's an understanding that goes beyond mathematics and embraces many other fields, those that are an authentic passion for him. Sometimes, around the table in a pizzeria, he talks to some of them about music, concerts, operas, and films. "Professor, in your opinion why did Einstein oppose quantum theory?" a boy asks him. "Because, according to Einstein, God doesn't play dice... And then, you see, the world of physical truths, like mathematical ones, is closed like a sphere. Every new vision, if it is profound, is an escape from this kind of prison. There may be some resistance to fleeing, otherwise you really can't see the reason why."

The exams resume. A lovely girl appears before him to face the terrible ordeal.

"Draw a straight line," Renato asks her with a distracted air, his right hand resting on his forehead as usual. The student draws a line. "Go on." The girl gets to the frame of the blackboard. "Keep going." The young woman comes to the wall, on which, disconsolate, she continues to draw a straight line, which as we know has no end. "Continue, continue." The girl gets to the door. "Go, go, continue outside. Goodbye."

In the evening, as often happens, Caccioppoli is having dinner with his friend Maria Del Re; in a corner, frowning, there is the student who was mistreated and humiliated in front of everyone in the afternoon. She is one of the professor's protégées, who Renato meets often, the delightful Raffaella Maione. "Lina," he addresses her in an innocent tone, "I see you are downhearted. What's the matter?"

29

The Battleship Potemkin
1952

It is June 16, 1952. On the steps of the university surrounded by police cordons, Caccioppoli gives a tough, worried speech. Two hundred students listen to him. He speak in favour of peace, but above all against the arrival in Naples of the new commander of the NATO forces in Europe, General Matthew B. Ridgway, the "plague man" accused of having used bacteriological weapons in Korea. And not only this. He launches invectives against the US marines lodged in a hotel opposite. His speech causes an uproar. At the end of the day, as he heads home, he is stopped by police officers. They take him to the police station, where they detain him for over 2 hours. He is accused of having organized the demonstration and stirring up the students.

Word that Caccioppoli was stopped and interrogated makes the rounds of the city, spreads to the university, branches of the PCI and the editorial offices of left-wing newspapers. Interviewed by *l'Unità*, the professor says with his typical sense of humour, but also with his stringent logic: "Having been taken to and received at the police station by some most affable officers, passages of a speech that I had given to the students were challenged with the utmost precision: a clear sign that strangers to the university had infiltrated my audience. And I was invited to take note of a warning not to 'promote' any demonstration not authorized by the police beforehand. Taking note, I declared, had it put down in writing and signed that, in rejecting this and any other intrusion into university affairs, I could not have discussed these if not with the Rector, my superior and equal: *primus inter pares*. Not having had the opportunity to meet the Rector during the event, I would not wish to have to conclude that the Rectorate is based in the political office of the police station.

L. Foschini, *The Friction of Life*, https://doi.org/10.1007/978-3-031-65262-2_29

The nonconformist and irreverent mathematician impatient of every form of authority, represents with his deeds not only an example, but also a unique opportunity for the city to emerge from the provincialism into which it has fallen. Caccioppoli, cosmopolitan, polyglot and at the same time so intensely Neapolitan, is a reference point for a group of young people who draw from the comparison with such an extreme personality an opportunity for growth and escape from the small world into which they were born and raised during the war and after the fall of fascism.

These are the years of Malcolm Lowry's *Under the Volcano* and of Sartre's existentialism. The drinking, the repudiation of bourgeois logic, the *nausea* of life that apparently have no meaning seem to be embodied in him, in his *maudit* aspect, in the rumours about his female loves, but also about his alleged, never confirmed, imaginary homosexuality. "We dressed à la Caccioppoli," recalls Raffaele La Capria, "open shirt, dark turtleneck, no jacket or tie, but a silk scarf around the neck."

One Sunday, at the end of a screening at the cinema in via Nisco, which as always he presented in front of an audience of enthusiastic *cinéphiles*, an air force officer approaches him. "The chaplain sent me," he says to Renato, evidently confusing the cinema club with a Catholic cinema club. "I would like to organize in Nisida (the small island where the Academy is based, author's note) a film show for the cadets, could you help me?" Caccioppoli appears enthusiastic about the proposal and promises to bring a beautiful film: "A story of navy ships, of the sea. An extraordinary thing," he assures. Satisfied, the officer leaves and Renato, wickedly assures his collaborators from the club who looked on in perplexity at the scene: "It will be a beautiful Sunday." And he adds: "We'll show them *The Battleship Potemkin*."

On the appointed day, the Academy room is packed with officer cadets and all the highest-ranking military men accompanied by their wives. There is great excitement as they await the screening. But as soon as the images of Eisenstein's film appear on the screen in close-up showing the grim, aroused faces of the Russian revolutionary sailors intent on hurling their own officers into the boilers, at first the audience is left dumbfounded with amazement, then they begin to protest furiously. All hell breaks out. People are screaming. The projection is suspended. Renato, with his friends Laura Sansone and Francesca Spada who accompanied him, beats a hasty retreat. The trio flee, amused by the stunt, along the bridge between the island and the mainland where a waiting taxi has been prudently arranged.

This story spreads like wildfire. It is talked about at the university, but also in schools where the attraction that Caccioppoli exercises on adolescents reaches the dimension of legend. I remember how much my sister yearned as a girl to meet him. At Umberto high school her classmates Annamaria Palermo, Raimonda Gaetani and Minni Minervini sing the praises to her of afternoons spent in the sumptuous home of Senator Mario Palermo with his daughter Gioia, the same age as them. Renato offers to give mathematics lessons to the very young and beautiful girls. Sitting on his hunkers on some steps in his friend's study, the professor unfolds metres and metres of toilet paper on which in his incomprehensible writing he lists an endless succession of numbers, square roots, algebraic fractions. The students don't understand a thing, but it doesn't matter. They can't take their eyes off that man who, enveloped in a spiral of scrolls of extra light white tissue, appears to them like an irresistible, fascinating, gesticulating ghost.

30

Words and Life
1951–1952

"Professor, are you a Marxist?" some comrade apparently asked him, receiving for his pains that ironic, half smothered laugh of his.

Caccioppoli never took a PCI membership card. The only badge that he wears is that of Picasso's dove of peace, but well hidden under the lapel of his raincoat. He takes part in marches, demonstrations for disarmament, and holds rallies, however in the face of the enthusiasm and sacred fervour of the militants he always remains ironic, doubtful, disenchanted. It is pointless to try to ascribe any affiliation to him. Ermanno Rea explains this well: "Caccioppoli is an instinctive type, someone who lives by his contradictions, with a critical relationship with respect to the party, experienced neither as a church nor as a hell. He is neither an assertive nor an orthodox type. He lives by his doubts." Yet, in the stubborn attempt to find a reason for his death, there are those who indicated the cause as the Soviet invasion of Hungary. The entry of the tanks into Budapest in '56—it was said—would have taken away all hope, to the point of seeing all his illusions vanish. Identifying, this time, his dramatic final decision no longer in the failure of a love, but in that of an ideal.

But there are those among his friends who maintain the exact opposite. For the jurist Francesco Guizzi, who frequented him between '52 and '55: "Suicide is always something indecipherable and there is never just one underlying cause. And it certainly didn't happen on account of the events in Hungary, quite the contrary. Renato Caccioppoli became reconciled with the party when the Soviet tanks invaded". Even though he had no respect for Khrushchev of whom he said, in his usual caustic, stinging way, to Mario Palermo: "When

L. Foschini, *The Friction of Life*, https://doi.org/10.1007/978-3-031-65262-2_30

the eternal father handed him intelligence he put a ladle full of shit instead of brains."

As for attitudes towards him, those of the diligent officials of the PCI, the most obtuse ones of course, are reminiscent in certain aspects of those of the police and the prefecture. They too look on the extravagant mathematician with frowning distrust. They can't stand his transgressive, derisive behaviour, out of any possible control, but at the same time they take advantage of his great influence over the young comrades.

Even though he knows him very well and is his friend, Giorgio Amendola tries to set limits on his impromptu speeches. And one evening, chatting amicably, he advises him not to talk off the cuff when he has to speak in public: "The list of points," he tells him, "is a sign of intellectual rigour and political discipline." The next day Caccioppoli has to give a speech before a crowded room. He arrives with a sheet of paper held between his fingers. But those able to take a look say that he is holding only a blank page. With the passing of time, however, even though he is an amazing conversationalist, words begin to get more and more rare, making way for that *horror vacui* that is Pascal's God-shaped vacuum. Not even words can fill that.

Renato shows the open palm of his hand to a student. "Look," he tells him, "that's the word." Then he points to his wrist: "And that's life." Then he moves his hand, in an impossible attempt to touch his wrist: "See? The hand gets close to the wrist, but it never grabs it. Same for the word. It barely skims life, but it doesn't grasp it."

In Bari, at the Piccinni theatre, the hall is packed with people who came to hear him on a topic dear to him, that of peace. As he is preparing to speak, he notices a grand piano on the stage. He approaches, sits down, and for the audience at first surprised and then enthralled, instead of a speech he holds a concert. The notes of Debussy's *Cloches à travers les feuilles* swell in the air.

31

A Pure Spirit
1951–1952

"Emma, go with Renato and call the lift." It is the winter of 1951. Caccioppoli is in Rome to meet Guido Castelnuovo, purged by fascism because he is Jewish, now senator for life and, like Renato, a member of the Accademia dei Lincei. The great mathematician is 86 years old and is seriously ill and will soon die. In his study there is also his daughter Emma, who met the Neapolitan professor several times, becoming more and more impressed by the mobility of his face, by the liveliness of the expression that continually shifts from happiness to sadness, from the mutability of his topics which vary from scientific seriousness to the most banal themes, always treated with the same verve. But this time the old senator looks at his younger colleague with concern and, when at the end of the visit, Caccioppoli takes his leave, worried, he tells his daughter to help Renato get into the lift. "He was afraid," recalls Emma Castelnuovo, "that he would fall down the stairs. He couldn't stand up. He was completely drunk." Once back at her father's side, she hears him mutter to himself: "I really don't think he's going to live very long."

More and more often, now, university colleagues, the scholars he meets at conferences, and the students he examines have learned to recognize the visible traces of alcohol in his staggering gait and they fear his sudden outbursts.

And yet, despite everything, even though deeply tarnished, the irresistible attraction that he arouses in the world around him remains intact. He can be sarcastic, cutting, sometimes cruel, he very rarely finds someone he finds stimulating to confront. It happens with his teacher Picone, with Miranda, with Cimmino and then, one day, also with a young man who, by no coincidence later turned out to be a great mathematician.

© The Author(s), under exclusive license to Springer Nature Switzerland AG 2024
L. Foschini, *The Friction of Life*, https://doi.org/10.1007/978-3-031-65262-2_31

He meets him during a seminar. Caccioppoli has just finished speaking and, while everyone is silent, he hears from the back of the room a shrill, sing-song voice and a nasal accent that, breaking the silence, begins to make a series of findings. Renato listens impatiently as usual, then to everyone's amazement he willingly offers the requested clarifications. But Ennio De Giorgi, that's his name, still isn't satisfied. he gets up again, insists, urges him: "Ah! And couldn't we say this instead? Couldn't it be done like this?" Caccioppoli, looking at him with eyes narrowed into slits, appears surprisingly happy with the reproof and the insistence and, before answering him, he comments aloud: "There is nothing more barbaric than a pure spirit." The silence becomes heavy. Everyone is tense in fear that something terribly caustic may come out of his mouth. And instead, he adds: "It seems to me that you are an exception."

"The topic of conversation," says Andrea Parlangeli, "was a problem about calculating the variation that Mario Picone called the 'ship problem'. It was about finding, among all the possible shapes of a ship's hull, the most convenient one from an economic point of view. The complexity arose from the fact that it was necessary to take into account two terms in 'competition' between themselves: the first was proportional to the volume of the hull and represented, in broad terms, the difference between the cost of the internal part and the achievable profit; The second term was proportional to the surface area and represented the cost of the external part of the ship. Picone was very interested in situations of this type and thought that Caccioppoli was the most suitable person to come up with a solution."

So, on his advice De Giorgi arrives in Naples to meet him, but this turns out to be no easy feat. Don Savino Coronato finds him lodgings with the Comboni Fathers and in the meantime works to get him a first appointment. Finally, after exhausting, patient work, he succeeds. Ennio and Renato begin to spend time together. They spend entire nights talking, tackling the issues. Sometimes it happens that in order to debate, Renato asks for a blackboard, surprising the young De Giorgi, because their meetings do not take place in a university classroom, but in a café, a bar, or a restaurant. The request is impossible to fulfil. Then Caccioppoli takes a sheet of white paper and says: "This is how things go," and draws two or three fairly cursory signs; he accompanies these with a concise but extremely profound speech, and in a few minutes he manages to give an idea of how things actually stand with respect to a certain problem. Parlangeli adds: "De Giorgi admires the master; he is fascinated by him. Renato Caccioppoli winks at the young man, but then takes to his heels, losing himself among the alleys of Naples, disappearing like a mirage."

In an article in *l'Unità*, De Giorgi talks about Renato many years later. He uses words full of respect for his life and his death, perhaps among the most lucid and profound things ever said about him. "As difficult and incautious as it is to enter into the mystery of a man," he writes, "if I had to see a thread between Caccioppoli's artistic, scientific, and social and civil interest, I would see the underlying aspiration to harmony, and the pain that all various disharmonies at various levels caused him."

Caccioppoli's relationship with Mauro Picone is very different from that with the young Ennio. The professor is his master, who believed in him right from their first meeting. Renato, while diverging from his master's political beliefs ("I was a fascist right from the start," Picone stated with pride in the 1930s), worked to ensure that he was not purged, precisely because he knew his great value and because he had a sincere feeling of friendship and affection for him.

One evening Picone, who has been teaching for years in Rome, visits his pupil at Palazzo Cellammare. They have an amiable conversation. At a certain point the professor gets up to go to the bathroom. He enters and is stunned by an absolutely unexpected sight. On the wall opposite the toilet, sheets of torn paper hang from a nail. Intrigued, he takes them and realizes, astonished, that they are recent articles of his that he had given to Caccioppoli for him to read in return for his opinion. He returns to the living room in disbelief, protesting vehemently, indignant. "Oh yes, you know," Renato replies calmly, with his typical childish cruelty, "you know that's where I go to concentrate."

32

Women
1954–1956

Women like Caccioppoli. He is charming, witty, and his reputation as a genius flatters the young women who seek his company. After Sara's departure he had many women, but none managed to take his wife's place. Despite his great success, his relationships with the opposite sex are difficult, verging on conflictual. Maybe that's why there is whispered talk of his homosexuality. But we know nothing about his male loves: if there were any and with whom, or if once again, as I believe, it is a matter of wholly invented rumours that feed the aura of legend that surrounds him. As for his women, however, we have several details.

Ornella Marzoli is a beautiful girl, with the allure of a model. With two marriages behind her, she is a Northerner afflicted by angst and a Sartrean nausea, quite common in those years among educated bourgeois youth. On account of her restlessness and certain family properties, including a pasta factory, Renato calls her "my doubtful miller". Together they spend hours drinking in bar Cristallo in Piazza dei Martiri, an elegant meeting place for the gilded youth of Naples, who they openly despised and made the butt of cruel jokes.

Then there is Luisa, petite but attractive. She attends all the professor's presentations at the cinema club and would even like to get engaged to him. In this regard, some cinéphiles see them discussing animatedly. Renato is explaining to her with a certain harshness that it is "absolutely out of the question".

Finally, the beautiful Renée. At the end of via Filangieri, on the corner with via dei Mille, at the fine Domus shop, in one of those ugly, grey Umbertine buildings that surprisingly contain splendid apartments furnished with paintings, furniture and sometimes fabulous porcelain, a likeable and enterprising

L. Foschini, *The Friction of Life*, https://doi.org/10.1007/978-3-031-65262-2_32

lady lived in the 50s, one of the many foreigners who arrived in Naples perhaps for love and then got caught in the city's dense web. Imprisoned by the fine weft of its enveloping threads, made up of exaggerated and slightly old-fashioned manners, of courtesies and hand-kissing, of sincere helpfulness that was cunning at the same time, with its indolent and tempting comforts. The top floor of the building is Renée's home from where she runs her dressmaking business frequented by elegant ladies from Chiaia and Posillipo. In the early days Caccioppoli was attracted by Madame's Parisian charm, which refers to everything he loves most in the world: France, its language, and its *esprit*. Renée speaks a strange and picturesque mixture of Italian, French and above all Neapolitan. She has a beautiful house on Capri, which she gets to with a motorboat that she drives personally, an exceptional fact for the time, especially in Southern Italy. If someone pointed out how much more prudent it would be to travel aboard a comfortable steamboat, she would promptly reply by rolling the r: "Better to die for the fish than the worms!"

For a fairly long period, between '54 and '56, almost every afternoon Renato leaves Palazzo Cellammare and takes the opposite route to the one taken in the morning. He turns right, walks along via Filangieri, and once he reaches via dei Mille, in front of Renée's street door, he takes the convenient stairs that take him to the top floor. What goes on between them is difficult to know for sure, but easy to guess. They see each other regularly, on a daily basis.

One afternoon, not finding her at home, the professor goes to look for her in the dressmaker's shop. For a few days it has been run by the young and kind Lucia Schettino, who has not yet had occasion to meet Caccioppoli. The young lady hears the doorbell, leaves the dress she is working on and goes to open it with a smile. She is expecting a customer. But the smile freezes on her face and turns into a grimace of disgust or fear. In front of her is a badly dressed man who, in a confused tone that strikes her as menacing, asks for the lady. Lucia gets the impression that he is even begging. His appearance is disturbing. The young woman becomes alarmed and yells all in one breath: "Go away. There is no one here. Away with you!" And pushing him onto the landing she slams the door in his face.

From the back of the room Madame asks her who it was. And, from the woman's description, she immediately understands who it was. She drops everything, storms out of the shop and runs down the stairs calling: "Renato! Renato!" She catches up with him as he is leaving the building. She flings herself at this slender twig of a man and holds him back by one arm. She apologizes, giving a flustered explanation for that unfortunate episode. She is afraid he will be offended and that he won't want to see her again. She expresses all of these fears in one breath and in more confused language than usual. He gives that sardonic laugh of his, but with an unexpected undertone of sweetness. A few days earlier, at Maria Del Re's house, to someone who had asked him what, in his opinion, was the most important phrase in history, he replied with the same words he now addresses to Renée to reassure her: "Fear not, my dear, the heart wants what the heart wants."

Paola Trapani, a former student, deserves a separate mention. She is the holiday companion in the Swiss Alps, the friend with whom Renato shares readings and amusements. She represents, it seems that Renato tells his friends, the *esprit de géométrie*, rather than that of *finesse*. Intelligent, bourgeois, she follows him with loving solicitude. She urges him to dress well and accompanies him to the shops in the city centre to buy suits, linen and silk shirts, and ties. Elegant apparel that will never see the light of day.

As soon as it is purchased, in fact, it ends up in the wardrobes or in the closets in the apartment in via Cellamare, and there it stays. Their affair ends at a certain point, but it is followed sometime later by an unexpected twist. One evening in a trattoria, while he is having dinner with friends, Caccioppoli suddenly comes out with this: "Have you heard the latest? My brother Ugo got married. But you'll never guess with whom. The bride is Paola!" The joking tone clashes with the expression on his face, on which—those present will say—a grimace of pained surprise appears. Caccioppoli is embittered, sad. This is one version of events. Another version from Ugo's and Paola's entourage, denies it years afterward. And, on the contrary, Renato is described as even amused by the singular coincidence.

Maybe one ought to have more elements to get to the bottom in this story, to understand why Ugo, an upright magistrate and fond brother, chose as his wife Paola Trapani, his brother's ex-girlfriend. The fact remains, and this causes some perturbation, that the date of the wedding, April 30, 1959, came one week before Renato's death.

33

Trafficker in Ideas
1953–1956

The more I delve into the folds of his life, the more I realize that, while the years of fascism were a painful and oppressive journey, those that followed the end of the war turn out to be even more bitter and tormented for Caccioppoli because they destroy all hope of freedom in him.

It is 1953. The Accademia dei Lincei awards him the President of the Republic's national prize for physical, mathematical and natural sciences. It is enlightening to read the rationale: "Some time ago Caccioppoli already attained a leading international position, among those few scholars of mathematics who actually make progress with discoveries that open up new possibilities and new horizons with the creation of fruitful research methods [...]

You can always be sure to find, where mathematics encounters serious difficulties, a contribution from Caccioppoli, useful to progress." The judging commission concludes: "Renato Caccioppoli, a great mathematician who dominates with impressive creative power the three fields of analysis, topological, real and complex, whose assiduous work as a scientist and as a teacher greatly honours our country."

These were the years in which his fame transcended Italian frontiers. Yet even though he is invited to numerous prestigious international conferences, his glittering university career is systematically held up, hampered by tight police monitoring, a watchful surveillance that has nothing to envy what he underwent in the 20 years of the fascist regime.

It is the summer of 1953; Renato is invited to a conference of mathematicians in Poland. He prepares to leave. He's in one of his rare serene states of mind, full of satisfaction, which only mathematics and music can give him. He has an important work to present. But day after day he finds himself

© The Author(s), under exclusive license to Springer Nature Switzerland AG 2024
L. Foschini, *The Friction of Life*, https://doi.org/10.1007/978-3-031-65262-2_33

having to face a thousand unexpected difficulties. In the end he is forced to give up. On August 11th he wrote to Mauro Picone: "Foolish police harassment leads me to renounce Poland. Just imagine, after weeks of foot dragging they returned, a passport… cancelled entirely (even for France!) but…extended for Poland and 'transit countries' (?) until the 6th of September, the opening day of the conference. With such a passport, I would hardly get any further than Tarvisio."

For months there was an extenuating tug of war between him and the police station which wore him down and, in the end, defeated him. He cannot leave. But he doesn't give up: "However, I am not giving up on presenting a communication to Conference. I have in hand one that should be first rate and I would like to entrust it to you." And it is moving to read with how much pride, but also with how much modesty, he asks Picone that the results achieved be presented to colleagues in the international community.

History repeats itself several times. One of the greatest mathematicians of the twentieth century is forced to submit to the humiliation of contacting the police to be able to leave Italy, always receiving an unyielding denial. It's July 1954. Another important conference is about to open in Amsterdam. Renato replies to Picone who asks him for information on his departure for Holland: "My dear Mauro, I wrote to you several months ago that I would 'not' have gone to Amsterdam and explained why. You replied saying 'I don't accept [this]' which I took in the right sense, that is, as an impulsive manifestation of your generous temperament, which no one, believe me, appreciates more than I.

But if anything, you should have said 'I don't accept' to the Scelbas, to the Fanfanis, or to whoever sees in many Italians, and in me among many others, if not exactly 'enemies of the homeland' at least 'discriminated' citizens, i.e. those who do not enjoy all constitutional rights."

He was, as always, refused a passport. And not only that. With subtle perfidy, pretending to suggest a way out of the country, diligent police officials recommend "trying" to ask for a permit for some days, with a mandatory itinerary. Caccioppoli defines this for what it is: "a kind of deportation order. The frontiers of our 'free' country," he writes bitterly, "can be crossed by a recognized drug trafficker, but not by professor Renato Caccioppoli, rightly or wrongly suspected of trafficking in ideas."

But his existence is not easy even when he moves within Italian borders. He has to leave for Varenna, where he is to hold a seminar on mathematical studies. A carabinieri sergeant shows up at the Institute of Analysis and questions Carlo Miranda. He asks him a lot of questions about Renato. "In the legitimate suspicion," comments Caccioppoli in a letter to Picone, "that subversive activities might take place on Lake Como."

How much do the continuous refusals, the tailing, the suspicions affect his ingenious, brilliant but fragile nervous system put to the test too much in the course of life?

In 1956, the mother ever present in his life, Sofia Bakunin dies. Ready to protect him with delicacy, stepping in to help him in the darkest moments that have seized him since he was a child, calling their friends, Palermo, Amendola, Scorza Dragoni, so that they rush to support him with tact.

Since his mother's death, his mathematical streak seems to have dried up and at the same time the already fragile health and the tenuous mental mechanism that moves his wonderful inner world decline precipitously. This is not a mere coincidence.

34

Francesca Spada
1957

"If he digs and digs," Renato explains to Renzo Lapiccirella, a friend and deputy editor-in-chief of *l'Unità*, ("one of the purest Marxists in Naples" as Anna Maria Ortese defines him), "in the end, what does he who has eyes to see discover, over and above the confines of disorder? He discovers fragments of harmony, of perfection. And he gives them to us. This is what geniuses do, be they mathematicians, musicians or poets."

Harmony. Throughout his life, Caccioppoli aimed to attain it as an attempt at synthesis, therefore peace, between his various restless, musical, scientific, and civil souls. Harmony is perfection. "Because in him," explains Ennio De Giorgi, "there is not the idea of disorder but rather of Pythagorean harmony, that is, the idea that in the end a truly interesting mathematical construction must be beautiful and harmonious, it cannot be a messy, incoherent construction, devoid of beauty."

And Renato seeks harmony, on the piano too, for hours, every day. For some time now he has also been playing four-handed pieces with a young woman who everyone describes as extremely charming. Francesca Spada, born in Tripoli, is not only beautiful, but she also has intelligence and temperament and is an elegant, refined pianist. She studied at the conservatory and had—connoisseurs note—"a noble touch, a natural sense of phrasing". She is Renzo Lapiccirella's partner, and she is the communist newspaper's music critic. Caccioppoli and Spada have many things in common. An existential ennui that pervades them, a good dose of non-conformism that in a woman immediately causes scandal, and above all a sensitivity, a malaise that both try to soothe in music.

L. Foschini, *The Friction of Life*, https://doi.org/10.1007/978-3-031-65262-2_34

Renato likes Francesca because, like Sara and more than Sara, she has all the characteristics that attract him: she is cultured, she is not provincial, with a strong character, unconventional. This doesn't go unnoticed. In the closed, largely moralistic environment of the Neapolitan PCI and the editorial staff of *l'Unità*, the understanding between her and the mathematician is noticed. Gossip flourishes. They meet at Palazzo Cellammare. They play Brahms together. Sometimes Joseph Haydn's *Variations on a Theme*. Every now and again they break off to drink, then continue until late at night, very often with friends, but sometimes also alone.

The rumours, at first a few words dropped in passing as if by chance, then more and more consistently, state assertively: "Caccioppoli and la Spada are having an affair."

Once again, as in the case of the fictitious episode when Renato and Sara allegedly performed the *Marseillaise*, we are faced with a well-constructed story, a sign that Caccioppoli's fame in the city's imagination must find outlet in fantasies and that these fantasies generate fake news. Gossip passes from mouth to mouth and becomes a certainty. "On a rainy evening Francesca and Renato," they say with morbid excitement, "were seen walking along Via Roma. The professor was wearing the inevitable Aquascutum and Francesca was also wearing a raincoat, except that hers was a transparent summer trench coat. Beneath, she was completely naked."

We don't know if there ever was, between these two neurotic, melancholy, enigmatic people, something more than an intense friendship. But 2 years after Caccioppoli's death, in 1961, Francesca Spada committed suicide. And Ornella Marzoli, "my doubtful miller" as Renato jokingly called her, was also to kill herself.

35

Giovanni Ansaldo
1957

In the middle of the night the editor-in-chief Giovanni Ansaldo dons his trilby, takes his stick and leaves *Il Mattino*. On the stairs he comes to the incredibly varied world of the Angiporto Galleria, where his newspaper has not only been based since the time of Edoardo Scarfoglio, but whose first floors are crowded with the Naples editorial offices of *l'Unità*, followed by those of *La Voce*, *Paese Sera*, *Tempo* and the Ansa news agency. He encounters comrades Mario Sansone, Fausto De Luca, Ruggero Guarini, Renzo Lapiccirella, the beautiful Francesca Spada and sometimes even Caccioppoli, on one of his visits to the communist newspaper. Usually, the professor brings a bottle of Stock Medicinal, a brandy in vogue at the time, to hand it round in the editorial office. Characters who often border on folklore come up and go down. In addition to journalists from different and opposing papers, "capable of ignoring each other to the point of insolence", there is a coming and going of printers, proofreaders, secretaries, visitors illustrious and otherwise, plus the flamboyant young ladies who work at the Buonanno boarding house, a renowned and popular brothel. When he passes them, Ansaldo responds to their smile by chivalrously tipping his hat.

The editor leaves the Galleria Umberto and finds himself in front of the majestic façade of San Carlo, thus passing from squalor to splendour in an instant, as can happen only in Naples. The route home is not long, but after a hectic day, long enough to mull over the events that have led him, 8 years ago now, to editorship of the daily. A task surprisingly entrusted to him, who in the course of his existence has experienced everything. From his friendship and collaboration with Piero Gobetti to joining the manifesto of the anti-fascist intellectuals set up by Croce, from prison in Como to confinement on

© The Author(s), under exclusive license to Springer Nature Switzerland AG 2024
L. Foschini, *The Friction of Life*, https://doi.org/10.1007/978-3-031-65262-2_35

Lipari, up to the "great leap" that led him in 1936 to edit *Il Telegrafo* di Livorno, owned by the Ciano family. Finally, the post-war period and prison on Procida, the isolation he avoids thanks to his friendship with Leo Longanesi, collaboration with *Il Borghese* and finally the unexpected proposal to direct the great newspaper owned by the Banco di Napoli, which however lies in the orbit of Silvio Gava's Christian Democrat party, with the aim of countering *Roma* owned by Achille Lauro.

Ansaldo is 62 years old, a considerable age for those days, and his past weighs heavy upon him, but this opportunity represents a shot at redemption and a challenge that he does not want to relinquish. Many thoughts run through his mind as he comes to Piazza San Ferdinando, takes via Chiaia and, before reaching Piazza Santa Caterina, turns right to take the steep street that runs alongside the former air raid shelter which has now become the mammoth Metropolitan cinema. "What an absurd city," he thinks, "which makes those who are not Neapolitan absurd, too! Indeed, it acts on him, transforming him into a grotesque thing." And his thoughts turn to Curzio Malaparte, who sued him for "misappropriation", because just after he arrived in the city as a guest of the bookseller Gaspare Casella in viale Calascione, he had slept on a mattress owned by Malaparte.

If there is a moon, even before tackling the hill Ansaldo looks up and sees the triple white, time-worn battlements of Palazzo Cellammare, which rises up with the magnificence of a castle. More than a century earlier it hosted the legendary glories of Prince Michele Imperiali di Francavilla and the famous landscape painter Jacob Philip Hackert, whom Goethe often visited when in Naples. Those who go up the steep little street leading up to the entrance gate do not immediately sense its grandeur, which manifests itself little by little, step by step. At first revealing the inaccessible and fortified aspect of its scarp walls in a creamy ashlar (marvellous walls that were unable—thinks the editor—to resist the onslaught of the populace in Masaniello's time), then as you get closer it reveals the profile of a fairytale castle with high walls of a pale, soothing pink, interrupted by large windows overlooking the sea.

Like every night, tonight too, even before reaching the dark baroque lava stone of Fuga's arch, surmounted by the coat of arms and closed by the heavy gate, Ansaldo stops to catch his breath. The access ramp runs around a flowerbed with slender palm trees which, in his opinion, give the building a vague hint of the East. But at that hour, which already heralds the first light of dawn, he is not the only one to tread those stones. From afar he glimpses a swaying figure ahead of him, slender and rakish, wrapped in a light trench coat, in summer as in winter, struggling to insert the key into the gate lock.

"Good evening professor!" Ansaldo thunders in his drawling Genoese accent and, on reaching him, he hastens to help him by quickly inserting his key in the keyhole, throwing the gate open. A "*grazie dirett*ò", mumbled and slurred by a mouth numbed by wine, sounds like a grunt. Then the man in the raincoat, twirling his slender bamboo stick, moves forward along the road, still uphill, that leads to his home. Ansaldo escorts him silently for the short journey, walking next to him, seldom exchanging a word, up to the small door that opens on the left under another arch. Discreetly he makes sure that the professor enters the small, damp, magical cavern where he lives, and then sets off slowly across the garden to the wing of the building where his own apartment is located.

He doesn't have time to walk away before notes are already spreading through the air. The professor isn't sleepy. He has sat down at the piano, and with his impeccable touch that no drunkenness can affect he is playing a poignant Chopin nocturne.

36

The Inane Attempt at Seriousness
1958

"Renato Caccioppoli belongs to legend." This is how a very young reporter begins by introducing him in an interview with her given by Renato at the beginning of 1958. It is included in a book, *Napoli e i napoletani,* a collection of statements made by the protagonists of the city. There's a bit of everything. Lawyers, entrepreneurs, politicians, aristocrats, good-for-nothings, comrades, Christian Democrats and monarchists. It is a truer portrait of the city than any treatise ever dedicated to it.

Why does Caccioppoli agree to become part of this bizarre congeries of characters, most of whom he thinks belong to the category of the "living dead"? I think he did it because Wanda Monaco, 18 years old, must have made an impression on him: not only is she very pretty, but she is also and above all sharp and intelligent.

What Wanda writes gives us an idea of the fame Renato enjoyed in his life among his fellow citizens: "Caccioppoli has always existed here and has never existed: coeval with every age and a citizen of the world. He himself seems to confirm his own surprising contradictions and the absurdity of these syllogisms. His mood swings are the result of a temperament whose nervous delicacy is directly proportional to the greatness of his genius. His whimsy, his habits, his examination systems are also talked about among those who have never known or met him—and who will never have the opportunity or reason to know or meet him—because they know that 'there is' a fabulous character called Renato Caccioppoli."

The answers that the professor gives to his interviewer are fun, relaxed, and, in his often hyperbolic and paradoxical style. However, a calm and pleasant helpfulness shine through, which is surprising, knowing how tormented his

© The Author(s), under exclusive license to Springer Nature Switzerland AG 2024
L. Foschini, *The Friction of Life*, https://doi.org/10.1007/978-3-031-65262-2_36

inner world was. A few witticisms also crop up here and there that seem to sound sincere, as when asked what essential trait of his personality is least noticeable and he replies: "The inane attempt at seriousness."

"If you had to change job, what would you choose?" Monaco asks. And he provocatively replies that were it not for a fear of overwhelming competition in Naples, he would opt for unemployment.

"What is your orientation towards women?"

"To orient myself I would need a compass, lost a long time ago, if ever owned."

With an obscure adverb that only he and a few others know, *onninamente*, absolutely, he declares he is completely in favour of the industrial development of Naples, but while waiting for the industries—he adds—the Neapolitans make shift for themselves.

"Do you want to tell us which circles you frequent?"

"The centre of the Gallery and the rotunda of the Caffè Gambrinus."

"Your hobby?"

"Maths, sometimes."

"Do you have any little idiosyncrasies?"

"Making fun of people, willingly accepting, of course a taste of my own medicine."

"What quality do you appreciate most in a man?"

"Manliness. We're Italians."

"And in a woman?"

"Chastity. We are vassals of the Vatican after all!"

"Who is the personage linked to the history of the city that you remember with greater fondness?"

"I am uncertain between Cesareo Console and Achille Lauro."

The very long interview ends like this: "Enamoured of Naples—he wouldn't know how to live anywhere else—the aspect of the city dearest to him is the one that is disappearing, because he says: 'I feel part of it.'"

This last sentence is almost a confession. The place where he was born no longer exists. Suffocated by concrete, neglect and speculation, it is dead forever. And he too is beginning to die.

37

Wanda Monaco
2020

I reach Wanda Monaco by phone in Stockholm, where she lives with her Swedish mathematician husband and where she carries on an interesting activity as an avant-garde director, actress and writer which also led her to work alongside Ingmar Bergman and his great actors, including Erland Josephson. She left Naples more than 40 years ago, but the account of her meeting with Renato Caccioppoli when she was just 18 is so vivid that it opens up unexpected glimpses of him and the last period of his life. The meeting with the Neapolitan mathematician—she confesses to me—was fundamental to her existence.

The young, already talented woman who collaborates with *Il Mattino*, was tasked with doing the interview which I have just reported in part. Right from the phone call she made to Renato to arrange an appointment, Wanda remembers lots of laughter. And this light-hearted joy is the fil rouge of a brief but intense friendship. They meet for the first time in Piazza Vittoria and from that moment an assiduous and joyful frequentation came into being with the Teatro San Carlo as its backdrop, where they start going to concerts together.

Renato understands how boring the girl might find Wagner and he confides in her, perhaps to please her, that he too is sometimes irked by him and, having the seat directly behind hers, during the heavier musical moments he kicks her seat in complicit solidarity. But what proves particularly intense occurs on their returns from the theatre of an evening, in a mild autumn, when Caccioppoli accompanies her from Piazza Plebiscito, walking along Via Caracciolo, to Viale Elena where Wanda lives. Despite the difference in age, the girl and the mature professor who talk animatedly as they walk along the sea front, which at that hour turns cobalt, have a lot in common. Starting

© The Author(s), under exclusive license to Springer Nature Switzerland AG 2024
L. Foschini, *The Friction of Life*, https://doi.org/10.1007/978-3-031-65262-2_37

with dress, he slovenly and she black and deliberately sloppy like her inspiration, Juliette Gréco, (these are the triumphant years of existentialism) down to her long unkempt hair. The pair are harmonically complementary. On those endless walks, Wanda Monaco was struck by Caccioppoli's feet which, she says, seemed to be double jointed and were the cause of his shambling gait.

The conversation between them is fluid, jocular and light-hearted, but also extraordinarily profound. The girl is won over by this man, by his intelligence and his originality: "There are people who have a particular diversity. He was different," she explains to me in theatrical language more suited to her, "which drove him to a need for self-expansion, not to try his hand at the limelight, but to receive something from the limelight. This was Renato. He didn't perform. He wanted to take something, from everything and everyone, but not the usual clichés."

Walking, conversing with him, stopping in the small bars along the road to sip countless coffees, the girl discovers that the professor, considered crazy and rebellious, seeks pleasure in things above all. An attitude that fascinates her, but which is deemed odd, suspicious, irregular, in the world around them, in the milieu in which they were both born. "In the culture of the Neapolitan bourgeoisie" Wanda explains "you could confuse Rossini with Pergolesi, but you could not break the rules, or make certain mistakes in certain behaviours." Renato does not deny these rules, but he has an existential, vital need to break them: "I have seen few people like Caccioppoli, who needed to have fun as if in a continuous search for pleasure." And in this regard Monaco finds a distinctive trait that unites Renato with pure mathematicians that she knows and, through her husband, has frequented for years: "It is a recurring characteristic of theirs to seek and find fun. Pure mathematicians are strange, they have a sense of play. The game of a mind that cannot sink fully into reality."

Listening to Wanda Monaco I think of the many pranks, the tricks that Renato loved to play and which at first I judged to be a childish characteristic of his temperament. But now perhaps I glimpse the origin of this. It is possible that Caccioppoli sensed the relationship with a dynamic and concrete reality with great discomfort, like a volatile relationship that could have continually called into question and destroyed his scientific research.

I read what Chiara Valerio writes about this: "The intention is not to say that science is opinion, but to introduce the advisability of a meta-science, something between introspection and a question of method [...] for progress in the scientific method that does not consider noise or uncertainty as a stumbling block in the inexhaustible attempt to describe and represent the world, but constitutive pillars. Uncertainty as an assumption of the scientific method, starting from that testing and testing again that is precisely method and never goal [...] As a person who has studied mathematics for a long time, I have always been fascinated by the attitude of those who, even knowing that they will only add one piece to the puzzle, construct methodologies whose coherence can be evaluated to attempt analysis and synthesis in the inconsistency and complexity of the world."

This is what Renato must have felt. In him, pleasure, play, are therefore a necessity. Once again Sciascia and his profound intuition come to my aid in describing Majorana's genius. Talking about a trick of his, a prank aimed at some colleagues, he writes: "This episode appears in the light of a profound 'superstitio', the kind that triggers neurosis: and mystification, theatricality and the prank are the precise counterpart of this—as in every neurosis."

Therefore, to put an end to perennial instability, the continuous restlessness that produces an inextricable knot of suffering, for the brilliant scientist the game lies in escape from this dramatic precariousness. For the first time I see with certainty that Renato's whole life was a vain, desperate search. A search for results, goals, discoveries, pleasure and even joy. Always reaching out to grab a truth he knew might be ephemeral, fleeting, susceptible to denial.

At a certain point, the thread on which this precarious equilibrium runs is irreparably broken.

38

Pucundria
1958

Caccioppoli has been alone for some time now in the Palazzo Cellammare apartment where at the time he was living there with Sara he already liked to imagine it was populated by ghosts. Even the lovely walks with the very young Wanda have become fewer. He continues to live his usual life. Every day he walks to the university, on Saturdays and Sundays he enlivens the cinema club, then dines with a few friends at Vini e Cucina run by Dolores, a faithful and protective friend more than a simple host, or at da Umberto where, to order fresh fish, he jokes with the waiters: "How many corpses do you have in the fridge?" He enjoys himself sometimes, surprising the customers, by singing French songs that he knows by heart. The women in his life are present, but these are increasingly fleeting encounters.

Del Re's house remains a safe haven. He has dinner there two or three times a week in the company of old friends. Francesco Villari was 8 years old at the time and after more than 60 years he still remembers it. He had been living there for some months with his mother waiting for his father, the great historian Rosario Villari, to arrive from Messina where he teaches to take up the position of central editor in chief at *Cronache Meridionali*. Francesco is small, yet he is intrigued by this man of whom he retains a very vivid memory. He watches him. Renato goes from a torrent of conversation to absolute silence, from biting irony to gloomy sadness. He plays the piano, stops and goes into the next room, then he returns, restless, suffering. He's fed up with everything. Even his friends. One evening he suddenly gets up and approaches little Francesco. Grabbing his arm, he says to him in a complicit whisper: "Come on, let's get out of here. I'll take you to the cinema. Let's go and see Sofia

© The Author(s), under exclusive license to Springer Nature Switzerland AG 2024
L. Foschini, *The Friction of Life*, https://doi.org/10.1007/978-3-031-65262-2_38

Loren." And taking the child by the arm, they head for the Arlecchino cinema in via Alabardieri. The scene was to be repeated often.

The liveliness of the conversation, the speed of the mathematical intuitions, the masterful piano performances, and the stinging irony preserve moments of masterful greatness. His assistants sometimes accompany him along the usual route from university to home. A mandatory stop at the Gambrinus, where Renato orders two half cognacs from the bartender. Professor Renato Fiorenza who, then very young, was amazed by this singular request, which Caccioppoli explains like this: "At the bar ½ plus ½ is more than 1, because a half cognac is always more than half of a serving."

In Fiorenza's accounts, as in those of Wanda Monaco, a portrait of Renato outside the stereotypes emerges. An admiring and touching portrait of such a complicated man, but one full of curiosity and the light-hearted pursuit of pleasure and fun. The former pupil explains this to me with a love and a devotion to his master that has remained intact over the decades: "Perhaps the insistence and exaltation of his predisposition to alcohol is due to the fact that it was an excellent excuse to talk and write about him," says Fiorenza. "Also because a discussion 'with the people' about the Schauder-Caccioppoli method or the Banach-Caccioppoli theorem of 'contractions' is not possible; the experts could do it, but obviously they too realize that very few would understand."

Yet, mysteriously, there is something so magnetic about him that he attracts even the simplest of people from every social background, who not only know nothing of mathematics but also of the novels and poems adored by this polyglot and extremely cultured genius. On leaving the house in the evening, on the corner of via Chiaia, the professor almost always finds a small group of loyal followers waiting for him. Some are workers of the gas company located right in front of Palazzo Cellammare. There is also an ENEL employee and a fellow who makes chair bottoms out of straw who has his shop in an alley behind Via Carlo Poerio. They accompany him wherever he goes, trotting alongside him. What is Renato talking about? And what do they make of what the professor says?

Fabrizia Ramondino seems to respond to this question of mine when she writes about him: "Those who managed to reconcile the figure of the great mathematician, the humanistic, anti-fascist and pro-communist intellectual, with that of the 'dissolute genius', were the humblest of Renato's friends; hosts and hostesses, concierges, beggars, poor relatives; the Dostoyevskian 'humiliated and offended', so at home in Naples. They were able to grasp simultaneously, without finding in this is an irredeemable logical contradiction—the heights of the spirit, the malaise of the soul, the suffering of Renato's body.

Honour therefore to the Neapolitan people, who still recognize the spiritual investiture of their own kings, even feeling them to be flesh of their flesh, in the sufferings of the body as in the essences of the soul—which in dialect have a single expression: '*pucundria*'."

It is not easy to explain the meaning of this Italian term, without resorting to long periphrases. *Pucundria* indicates a state of mind with indefinite contours, a painful sorrow that comes close to melancholy but also drags along with it boredom, dissatisfaction and loneliness.

Lately something in him at times seems to be breaking down, stuck, blocked, numb, his gait is more uncertain, his lightning-fast perceptions deteriorate at some points, the flashing brightness of his gaze clouds over. Renato has even become more intolerant. He reacts violently if anyone enters the room where he is and starts talking in commonplaces. He has the rare ability to grasp the nature of the person in front of him. He feels hurt by their stupidity. He is not well. He begins to get agitated, until the hapless visitor decides to leave.

The Neapolitans observe him with solicitous anxiety as he walks the streets of the city. He wanders around stooped, prostrate, his step hesitant, without looking either ahead or around him, with half-closed eyes, the cigarette unlit in the corner of his mouth. Sometimes he gestures as if he were waving his hand to some musical theme, or he mentally pursues thematic calculations that he then writes down quickly, stopping for a moment, on a strip of paper torn from an advertising poster, promptly slipped it into a pocket where it will remain for weeks.

These are the days in which Caccioppoli gives presents to Fabrizia Ramondino's aunt, Mimisa Vico, *The Treason of the Intellectuals* by Julien Benda, the philosopher who passed away in 1956 who was also an aspiring pure mathematician, a career he later abandoned to dedicate himself to history. Like Sciascia, Benda too glimpses at the end of the scientific and human path of the intellectual, but particularly of the genius, a truth, a harmony that inevitably takes him beyond life. To a "vanishing point".

Renato spends his long nights wandering aimlessly. He goes to the Quartieri Spagnoli, enters the alleys of the city which, porous with tuff, change in those dark hours from grey to a mournful black. He carries his slender bamboo stick, which he often hooks over his shoulder or twirls in the air. From the slums, figures of women, often monstrous, look out, the same ones that Ernst Bloch saw: "horrible as the Graeae, as witches at the stake, but with an imposing and pious appearance." He slips into the side streets, climbs dirty, narrow and steep stairs in search of the taverns, the wine shops, the bars where he invariably holds a glass of wine between his fingers and recites in French to his drinking companions, porters, thieves, whores or some *femminiello* who don't understand and a couple of faithful students who follow him everywhere, Baudelaire's lines: "Il faut être toujours ivre, tout est là; c'est l'unique. Pour ne pas sentir l'horrible fardeau du temps qui brise vos épaules et vous penche vers la terre, il faut vous enivrer sans trêve."[1]

Sometimes he lets himself collapse, tired, on the steps of a church. He sits there for hours as if asleep. One night one of his students, who would later become known for his verve and great ability as a populariser, Luciano De Crescenzo, recognizes him and, seeing him slumped there, thinks he has fallen ill. He approaches him solicitously: "Professor, do you need help?" In response, Renato beckons at him to sit down next to him and then says, in a completely unexpected way, as if the words arose from some intimate reflection of his: "When something scares you, take measures. You will realize that in the end it is a very small thing."

As the days pass, however, it is difficult for Renato to cope with his fears, with the increasingly menacing torments that invade his mind. These are the evenings in which Gustaw Herling comes across him, describing him with "the thin, tormented, almost bird-like face, a curl over his forehead. Every so often he would stop, take out a packet of cigarettes and, leaning against the wall, write on it." To the philosopher, Caccioppoli appears as "a walking monument to solitude and isolation. While I was following him," Lidia Croce's husband writes, "I overtook him, I observed him from the other side of the road, I felt something like the attraction of a mirror, but without having the courage to get too close to that dark surface."

---

[1] You must always be drunk; that's all; that's the only problem. Not to feel the horrible burden of time that breaks your back and grinds you into the earth, you must get drunk without respite.

39

Caccioppoli's Apartment
1958

With each passing day, the charming apartment in Palazzo Cellammare becomes more unadorned, as if Renato were slowly emptying it. Friends, colleagues, some students lucky enough to be received, notice this.

Maurizio Valenzi, who has frequented the house since the tragic night of June 11, 1946, when the headquarters of the PCI in via Medina was stormed by royalist marchers, is amazed by this. These are the crucial hours in which we await the official proclamation of the results of the referendum and the victory of the Republic. The clash causes the death of almost 10 people, all very young, and about 50 injured. In the general stampede someone shouts at Valenzi to join him at Caccioppoli's house. Maurizio arrives and already finds many comrades. There are Amendola, Sereni, Palermo, Cosenza, Alicata, the owner of the house and his wife Sara. From that day Renato and the future mayor of Naples begin to frequent each other, becoming friends. But, for Valenzi, in recent months something has changed in Caccioppoli. He has the feeling that in his nocturnal wandering his friend has lost all connection not only with space, but also with time. More and more often he shows up at Valenzi's home in via Toma, without warning, at the most absurd hours. Caccioppoli rings, wakes up the children who have been asleep for hours. Valenzi opens up and gently says: "Renato, it's after 11. It's nighttime." To avoid these inconveniences, he prefers to go and see the mathematician. But the last time he visits Palazzo Cellammare, upon entering the house, he is left speechless.

The apartment is desolately empty. Alone, in the corner, the black Petrof piano stands mournfully, next to it at the window the desk on which two photographs sit and, piled up in disorder, several sheets of paper with dozens

L. Foschini, *The Friction of Life*, https://doi.org/10.1007/978-3-031-65262-2_39

of formulae, absolutely indecipherable for those unfamiliar with the subject of his meditations. In another equally bare room, there is the bed. Books and newspapers are stacked on the floor of the living room where those that have not found a place on the table squat on the old sofa or on one of the two chairs that survived the clear out. A bottle of tremendous Australian White Swan rum stands on a stool. If a guest's eye lands on an object that mysteriously still remains, the houseowner notices and tells him quickly: "Do you want it? Take it away." What is Renato getting ready to do?, Maurizio wonders with a shiver.

40

The Two Photographs
1958

None of the two photographs lying on Caccioppoli's desk are of "grandfather" Bakunin. They are portraits of two beardless faces. The still childish face of the mathematician Évariste Galois, who died at the age of 20, and one who resembles a girl, a very young Arthur Rimbaud. I wonder why he arranged them on full display like this. What message does Renato want to send to anyone who enters his living room? The impression of many visitors on seeing those two old images is the immediate identification of the houseowner with them. But it's not that simple. What are, I wonder, the actual points in common that the professor shares with the two French teenagers, rebellious geniuses, present like domestic *Lares* in his apartment in Palazzo Cellammare?

Revolutionary anarchist, precocious genius, sentenced to prison for defaming King Louis Philippe, Galois certainly shared with Caccioppoli an early talent for mathematics, speed of perception and absolute speed in reaching the solution. Perhaps in moments of great suffering Renato envied his ability to meet his end so young, in a duel, uttering the touching phrase addressed to his brother shortly before dying: "Don't cry! I need all my courage to die at 20." There is something in common with Rimbaud, however, much more than restlessness, finding oblivion in alcohol, or the train ticket that takes him to Paris where he is accused of vagrancy, an experience similar to that of Caccioppoli's journey from Padua to Milan; the relationship with Arthur is something so profound that, upon reflection, it can reveal a lot about Caccioppoli's soul.

The photograph on the desk shows the poet between the ages of 16 and 17 as Verlaine describes him: "Tall, with the perfectly oval face of an angel in exile, disturbing eyes of a pale blue." It is the era in which Rimbaud affirms that "no one is very serious when he is 17". And Caccioppoli seems to reply to him when those who ask him which trait of his personality is the least noticed, he replies: "The inane attempt at seriousness."

Giuseppe Marcenaro, an author who explored the poet's soul with his usual acuity, explains to me: "Rimbaud fought against the demon that stirred within him. He opened all hearts where all wines flow." Already in 1871 Rimbaud wrote: "It is about arriving at the unknown through excess in all the senses... It is false to say: I think; one should say: I am thought. I am another." Words that shed light on the reason why the portrait of the boy from Charleville found its place on the desk of the mathematician from Naples. It is a viaticum that accompanies him in his tormented, desperate existence and the vain attempt to escape from himself, since for him mathematics, like music, are the only means that allow him to do so. But until when?

Renato very quickly reached the absolute awareness that the perception of everything inevitably leads to nothing. Hence the need, as it was for Rimbaud, to escape from himself. Playing the role of a vagabond wasn't enough to appease the sense of emptiness that precedes a desire for death. The realization that we have reached the point of no return is too strong.

41

The Friction of Life
October 1958

Towards the end of the 1950s many of Caccioppoli's friends, regular guests at Palazzo Cellammare, drinking companions, privileged interlocutors of his torrential conversations, or even simple acquaintances with whom he exchanged a few words, have now left. Naples empties. Journalists, writers, artists. Antonio Ghirelli, Giuseppe Patroni Griffi, but also Tommaso Giglio and Franco Rosi leave the city. Anna Maria Ortese and Giorgio Bassani have already left some time before. Then Pasquale Prunas, Ruggero Guarini. Everyone leaves, some to Rome and some to Milan. There is something that drives them out. A need to free themselves, similar to that which animates teenagers who want to leave the parental home. To step away from that caring love with which Naples envelops them and which irremediably leads them to a sort of infantile regression, to find their raison d'être elsewhere.

But there is a deeper reason pushing them to such a painful exodus, and it is linked to the very nature of the city: "Everyone had fallen here," writes Anna Maria Ortese of the Neapolitan intellectuals of those years, "those who had wished to think or act, all the languages had become confused and had increased the painful human vegetation. This nature could not tolerate human reason, and faced with humankind she moved her armies of clouds, of enchantments, so that man was stunned and submerged." Only a clean break can therefore cause an awakening from the ecstatic befuddlement. But it is a violent caesura that leaves those who find the strength to carry it out, to abandon Naples forever, irreparably mortally wounded, as Raffaele La Capria was to write only a couple of years later.

Renato suddenly realizes this flight and, despite affecting indifference, he feels the strong bite of loneliness, but he no longer leaves the city. Unlike

L. Foschini, *The Friction of Life*, https://doi.org/10.1007/978-3-031-65262-2_41

many others he doesn't hate or love Naples. His is a relationship that goes beyond love, it is a symbiotic relationship that generates an incurable sorrow for what the city could have been and never will be. Exactly what he thinks of himself.

And so he lets himself live. In the summer he gives up the rare trips to Ischia and Capri, where he also spent pleasant moments with Sara, or on Procida, where with Alberto Moravia and Elsa Morante he spent the evenings drinking and conversing in the small, enchanting Eldorado hotel. Even meetings with his poet friends, Éluard and Neruda, are now a distant memory. Every morning he takes the usual road again. His hatchet face, his gaze lucid and already vague, his gait gradually more uncertain, the brief stops to catch his breath, the grey woollen scarf wrapped around his neck; the return around 1 o' clock, a stop at the usual bar in via Toledo, re-entering the increasingly empty house where he continues to drink until he drops.

In August once again, for the umpteenth time in his life, he is subjected to a police check for "participating," he says in a letter, "in the peaceful demonstration in via Roma which had caused the riot police to land the usual hail of blows with their truncheons, without warning and equitably distributed among the demonstrators and simple passers-by". As for the rest, he carries on, drags himself along. On 7 October 1958 Carlo Miranda wrote to Mauro Picone: "Renato is fairly well even though he keeps on saying that he is no longer good for anything."

The International Writers' Conference is underway in Naples from 18 to 21 October. Paola Masino, partner of Massimo Bontempelli and author of notable books, takes part in the proceedings. On the 22nd, instead of returning to Rome, she decides to stay for another day. She wants to see her Neapolitan friends. Invited to breakfast with her friend Maria Luisa Costantini known as Mimma, wife of the engineer Sante Astaldi, she meets Renato. The long hours they spend together leave her shocked. Caccioppoli opens up to her with a dramatic sincerity that, I believe, he has never done with anyone. Masino is so moved that, as soon as she returns home, she feels the need to get things off her chest, to report what she has experienced, and she does so by writing to her mother. Perhaps out of delicacy, out of discretion, or because she is so shaken that she seeks comfort in maternal warmth. Here's what she tells her, like a great writer, in an anguished and touching way.

> Dear mum,
>   here I am back in Rome.
>   I will tell you about the Naples Congress when I see you. […]
>   I saw Caccioppoli again, who is now truly a skeleton.

I feel sorry for him and I wouldn't be surprised if he killed himself.

He offered to lend me a million—he says he is very rich and he doesn't spend any-thing—but, as usual, I didn't accept. I should have had more resistance and stay with him all day but, after five hours of his company, I felt as exhausted as I do after five months with Massimo.

Because you have to respond to him and respond as he wishes,

otherwise he gets angry, offended, withdraws into himself. He is an exceptional man, who cannot resist the friction of life, and who no longer does anything to live.

At table (he came with me to lunch at Mimma's) he only ate three grapes. If only if I had the courage to assume responsibility for his life, maybe he would save himself (I say this to mean any person willing to give him affection).

He is horribly lonely, he sent the maid away, he makes no phone calls, doesn't open the door for anyone. And what's more he disgusts himself.

"I wouldn't be surprised if he killed himself," Paola Masino writes. And again: "he can't resist the friction of life", using a term dear to her. Friction: resistance that a body encounters in its motion relative to another body. Wear and tear.

42

Caflisch
April 1959

Towards the end of April 1959, on one of those spring days that seems to me that only Naples can offer, when the air is enchantingly warm without being annoyingly hot and a light breeze comes from the sea, my mother leaves the house as she does every morning, crossing via dei Mille to *fare dei servizi*, an expression that in the city lexicon means to go shopping. She too is a bizarre character. Equally divided between the work of an elegant lady and a passion inherited from her astronomer grandfather, Francesco, for the stars. On summer nights she spends hours at the telescope studying the sky, and with equal enthusiasm she loves clothes and everything sugary that can be eaten, with unsuspected greed for such a slender woman.

Having arrived at the corner of via Filangieri and via Chiaia she goes in to Portolano, a large shop where Neapolitan ladies usually buy hand-sewn gloves made by women who live and work in the many slums of the Sanità district, but also fabrics that the seamstresses will then sew based on the models arrived from Paris.

She lingers for a while, unsure of what to do. She is looking for a fabric for a 7/8 suit that was very fashionable that year, but the colours are drab and the fabric has an overly thick grain that she fears, being so thin, might make her look fat. She leaves the shop and after a few steps she arrives in front of Caflisch. In the window, cassatas, *sfogliatelle*, but above all *scazzette del Cardinale* which herald the coming May. They are pastries, covered in a shocking pink icing, pyramid-shaped, filled with Chantilly cream and wild strawberries, which tempt her irresistibly. She lingers, then she enters. She's always in a hurry, but she is so greedy that she can't help herself, so she buys the sweet, but she doesn't eat it inside the pastry shop as a lady should, but on the

L. Foschini, *The Friction of Life*, https://doi.org/10.1007/978-3-031-65262-2_42

street, not to waste time. She comes out like this after a few minutes. In the end she has opted for a chocolate choux that she holds in one hand and, absolutely reprehensibly, she eats it while walking.

All intent on not getting dirty, she doesn't notice that in the opposite direction, coming from via Chiaia, someone equally distracted is coming towards her and she doesn't have time to step aside. So she collides with him, and the chocolate pastry in her hand smears a worn Aquascutum. She looks up and, very shy as she is, she wishes the ground would swallow her when she notices that she has fallen into the arms, one in a cast perhaps after a recent fall, of Renato Caccioppoli. With his good arm, he who is not exactly steady on his legs still manages to support her and prevent her from falling. He recognizes her, she smiles and with a look full of curiosity he asks her: "Are you Lorenzo's daughter?" My mother babbles that yes, her father is Lorenzo Caccioppoli. She would like to add something more, a friendly word, but she is too shy. He too would like to talk, but he is not encouraged by my mother's silence. He says something to her she can't grasp, in a kindly, almost affectionate tone, she was to tell me later. They say goodbye, who knows, perhaps intending to meet in more comfortable circumstances, and each goes their own way.

43

How Beautiful *a' Muntagna* Is Tonight
April 1959

April 30, 1959, Renato is seen for the last time at the university. On May 2 or 3 the mathematician Salvatore Rionero, who is waiting for his fiancée in front of the Gambrinus café, sees him coming down via Chiaia and emerging in piazza San Ferdinando. On looking at him he is flooded with a sense of desolate amazement, of indescribable melancholy. Caccioppoli seems barely able to stand. He walks with difficulty, hesitant, his arm in a cast at his neck. His proverbial thinness has reached "spectral" levels. His body appears as fragile as glass and gives the impression that, like glass, it may shatter at any moment. Rionero approaches him and, affectionate and deferential, he asks him: "Professor, how are you?" And Renato, giving Rionero his hand: "Rionero, how do you think I am?"

That day or the day immediately after Renato is with some students in a pizzeria in Posillipo. It's a warm and fragrant evening. They are having dinner in the garden overlooking a small beach and the moonlit gulf. In the background, beyond the sea, stands the dark mass of Vesuvius that the Neapolitans call *'a Muntagna*. At table one of the boys interrupts the conversation that revolves around the usual university issues and tells of one of their classmates who attempted suicide. "He took some pills," he says, "But he was saved." Caccioppoli intervenes: "He's a fool." His tone is contemptuous. He continues: "To really kill yourself, you do it like this: you take a gun, lie down on the sofa, put a pillow under your head to collect the blood, aim at the nape of the neck, which is more vulnerable, and then you shoot. In this way, if you really want to get it over with, it's impossible to go wrong."

In the embalmed air of the night, the intense, stupefying smell of mint and basil blends with the equally powerful saline odour rising up from the sea. All

L. Foschini, *The Friction of Life*, https://doi.org/10.1007/978-3-031-65262-2_43

133

around there ring out the notes of a song that a valet is singing while accompanying himself on the guitar. The poignant melancholy of the words softens the crudeness of those just pronounced by Renato:

How beautiful the mountain is tonight!
I have never seen it as beautiful as this.
It seems a resigned and weary soul
under the cover of this white moon.

44

What No One Knows
May 8, 1959

On May 8, 1959, after five in the afternoon, Mercedes (the name is imaginary, but the girl really existed and I met her) leaves the house and from Via Calabritto she sets off on foot towards Chiaia. She is 18 years old, tall, blonde, with dark eyes. She is wearing a light skirt, a white blouse and a scarf around her neck. On her feet are hand-sewn sandals bought in Positano. Under her arm she carries a couple of books and a maths workbook tied with an elastic band. Anyone who sees her might easily think that she is a student going for lessons in a subject she is not skilled in. In fact, there are just over 2 months left before the feared exams for the high school leaving certificate. The girl walks quickly. She quickly turns the corner of Santa Caterina and almost runs up the hill that leads to Palazzo Cellammare. She seems to want to pull out all the stops to get to where she is headed as soon as possible. A desire that is not at all plausible if all that awaits her is only a mathematics lesson. Actually, for some months she has often walked the same route in a state of mind of impatient anticipation mixed with slight anxiety. What has happened to her is a secret. Because few could understand. Mercedes is living through a completely unrepeatable story with a man, one of her father's friends, who is 40 years older than her and who she has known since she was a child.

It is difficult to understand the nature of this connection. It is a platonic love, perhaps, but no less intense, no less intriguing. Mercedes is overwhelmed by the unthinkable experience of being next to a person like him, listening to him, having fun, thrilled by the absolute awareness of experiencing the privilege of having captured the attention of a genius. She too, like her peer Wanda, is a girl of the Neapolitan bourgeoisie, her father is a well-known and influential professional, and she, like my sister, is profoundly fascinated by this man who is so different from everyone else, albeit fragile, worn out, physically exhausted, he reveals nothing to her of his intimate inner tragedy and, on the contrary, drags her along into a world a thousand light years away from the conventional one in which she grew up, opening up for her boundless expanses of emotions, discoveries, and startling things.

Mercedes isn't thinking about any of this, but runs. She wants to reach the place where she can experience these emotions as soon as possible. She goes through the second arch and finds herself in front of the little door. It's just after half past five in the afternoon. She rings. Once, twice, several times. Nobody answers. She knows that high up, in a hollow in the door, there is always a key. She raises her arm and feeling with her hand, she quickly finds it. She grasps it, slips it into the keyhole and, thinking that the person living in the apartment hasn't heard the doorbell because asleep, with little, muffled steps she enters the living room. In the half-light she walks slowly, even though she knows every corner of that room. She sees him resting on the sofa, but in a strange posture. She turns on the light and remains silent, astonished, because what she sees is too big, insurmountable.

The man lying supine on the pillow appears to be sleeping. But a blood-stain, which stands out, its vivid colour still fresh on the white linen pillow, and others on the floor reveal all the horror of what has happened.

A cold calm descends on her. She switches off the light, crosses the room and closes the door behind her and on what she has seen, as if she had never had that experience. Only a few minutes have gone by and she is already outside the apartment. She puts the key clutched in her fist back in the hollow at the top of the door. She goes back down the hill with light steps that become faster and faster, until she comes to piazza dei Martiri almost at a run and then her house.

Mercedes' life was to be forever marked by what happened that day and by the questions she has been asking herself ever since and to which she finds no answer: was Renato still alive when she entered the room? Could she have saved him? Why did she flee so quickly? Out of fear of death or rather out of the scandal that would ensue?

In the years to come Mercedes was to have other affairs, but she never bonded with anyone. She did not get married; she did not have children and she was to spend her life first with her elderly parents and then in melancholy solitude. Within her, for years and years, the almost adolescent love and the tragic epilogue that in an instant made her become painfully grown-up were to remain buried, in a doleful process of displacement.

Only much later, in her 50s, was she to find the courage to tell her cousin Marta what happened to her. Marta (this name is imaginary, too) being the only person, given her relentless craving for freedom, who in her opinion might understand. And Marta, shocked by this revelation, told it to me because she knew how attached I was to Renato.

I have tried several times over the years to return to the subject with her. I asked her to let me see Mercedes again or at least to delve deeper into what she had told her, but she, usually so adorably talkative, then fell into absolute silence. Now Mercedes is dead and Marta is gone too, but I decided to add this last short piece to the mosaic of Caccioppoli's life.

45

What Everyone Knows
May 8, 1959

On May 8, 1959, Renato killed himself exactly like this, with a shot fired from a Beretta 7.65.

The previous day he had been seen in Via Chiaia, where he had probably gone to the bank to get the gun from the safe-deposit box. His lifeless body, supine, on the sofa, was found around 5.45 pm by the maid, Assunta Russo, 49 years old, living in via San Rocco in Ponticelli, who had been working for him for about 19 years. Not much blood on the pillow, next to a cup of tea and some breadsticks. The bullet that exited his forehead was lodged in a shelf.

The report on the front page of the following day's issue of *Roma* reads: "Regarding the exact time the tragic episode occurred, there is reason to believe that the death of the illustrious mathematician occurred around 5.30 pm. In fact, when the body was discovered, the blood on the pillow, which had also formed a stain on the floor, next to the sofa, on a level with the suicide's head, was still fresh."

L. Foschini, *The Friction of Life*, https://doi.org/10.1007/978-3-031-65262-2_45

46

Truly Orphans
May 9, 1959

On May 9th the weather suddenly changed. After the beautiful previous days, so sweet and warm with sunshine, Naples appears shrouded in a dark enchantment. A thin rain falls continuously over the city and the waves of the sea, swollen and livid, the same grey colour as the sky, crash violently on the rocks on Via Caracciolo.

Since the early hours of the morning a long line of people climb the slope that leads to Palazzo Cellammare, regardless of the water falling and forming large puddles on the bumpy pavement that leads to Fuga's arch where the gate is wide open. There are dozens and dozens of them, patient and silent, waiting to slip up the narrow staircase that leads to Renato's apartment. Some reach the bedroom where he rests, to say a final farewell, while others, since the house is too small to contain them all, stop in the large courtyard, exchanging bewildered glances and a few, disjointed words.

That Saturday Giuseppe Scorza Dragoni, talented mathematician and dearest friend, seeing that small crowd so silent and lost brings to mind the words of Phaedo's words to Echecrates on the last moments of Socrates' life, when Phaedo returns to his thoughts on his and "our misfortune. How great it was," he reflects. "We knew well that for the rest of our lives we would be deprived of one like our father, truly orphans."

Perturbed and moved, the professor enters the apartment. He sees the desk left unusually tidy and uncluttered. On the desktop only two or three sheets, arranged in good order, one on top of the other, with about ten formulae, at first absolutely indecipherable. He can't help but wonder if those last lines represent a desperate attempt on Renato's part to test his great ingenuity, to see if he was still capable of reaching the heights in mathematics he was

© The Author(s), under exclusive license to Springer Nature Switzerland AG 2024
L. Foschini, *The Friction of Life*, https://doi.org/10.1007/978-3-031-65262-2_46

accustomed to. And he remembers a theory of his, which he judges admirable and profound, of pseudo-analytical surfaces and of pseudoconformal representations of Riemannian surfaces, and another about the integration and search for primitives with respect to any continuous function. The last of Caccioppoli's scientific constructions.

At 4 o'clock on an afternoon that became unusually cold, under the rain that became even more intense, the funeral cortège leaves the house and reaches via Chiaia. A stream of people follow the coffin carried on the shoulders of his students. The many black umbrellas open to shelter from the water appear like a long mournful cloak, a funereal drape spread over the entire procession. At the head, among a sea of men, a small female figure stands out. She is Renato's cousin, Giovanna Bakunin, daughter of Carlo, faithful guardian of her aunt Marussia. The professor didn't come, she was too old and ill, they saved her the pain of telling her about her beloved nephew's end. She died less than a year later.

Students from the engineering course, but also from other faculties, parade along the street. We recognize professors Miranda, Carrelli, Fiorenza, Scorza Dragoni, among them the black cassock of Don Savino Coronato, Caccioppoli's "shadow". Rector Pontieri, with the entire academic senate, the academics of the Lincei, the communist senators among whom the heartbroken faces of Palermo and Valenzi, and the secretary of the federation Alinovi, can be recognised. Togliatti sent a telegram of condolences for "the passing of the illustrious scientist, anti-fascist, noble fighter for peace and democracy".

Next to them the socialist Francesco De Martino and the republican Francesco Compagna with some journalists of his prestigious magazine *Nord e Sud*. In a bizarre and casual mix, walking side by side are the artisans from the workshops in the alleys of Chiaia, some workers from Italsider and San Giovanni a Teduccio and the employees of the gas company. Next to them are the musicians of the conservatory, enthusiastic music lovers, including aristocrats such as prince d'Avalos and the Marchese Parisio Perrotti. Confused among the crowd, the writer Luigi Incoronato who, upon returning home, dashed off these verses in one go: "Your funeral has passed / in our heart / like the irrationality / of a love too difficult / of rationality." Six years later, Incoronato, too, committed suicide.

The chaotic traffic that clogs the city at that hour suddenly disappears. And the people, silent, line the streets the professor walked every day. Traders come out of their shops, and from the windows and balconies of the beautiful buildings in Piazza Santa Caterina and Piazza dei Martiri many look out to accompany the coffin with their gaze as it heads for the Poggioreale cemetery. It is not simple curiosity. It is an intensely lived participation, a mass of collective pain that suddenly and unexpectedly reveals how the life and death of Renato

Caccioppoli are profoundly intertwined, through some mysterious alchemy, with the deep roots of the city.

In the afternoon my mother hastily said goodbye to us and left the house mentioning some urgent *servizio* to attend to. At a fast pace she walked along Via dei Mille, via Filangieri, vico Alabardieri, piazzetta Rodinò, then she took via Carlo Poerio and came out on the corner with Piazza dei Martiri where the Salus pharmacy is located. There, leaning against the frame of the entrance door, she witnessed the passage of the hearse (Fig. 46.1). In those moments, as was to happen to her many, many years later, she feels a strong, pungent, regret for her damned insecurity. For the words unspoken in the unforgettable, fleeting encounter of a few days earlier. And she sees his gaze again, dark, liquid, sweetly inquiring, which only briefly met hers, shy and blue, opening up infinite spaces of intelligence.

Fig. 46.1 Naples. Funeral of Renato Caccioppoli

47

Conclusion
Family Considerations

Ettore Perozzi, the only one in our family who dedicated himself to the study of the stars and planets observes: "There is a parallel between the 'life' of an asteroid and that of Renato. An asteroid may remain in orbit for much of its life, hundreds of millions of years between Mars and Jupiter and then suddenly show signs of instability. Its orbit then begins to wobble fearfully, and within a short time (astronomically speaking) its eccentricity will bring it to intersect the orbits of Mars, Venus, Earth, up until the inevitable epilogue: crashing onto their surfaces or ending up straight in the sun, melting in its immense furnace."

L. Foschini, *The Friction of Life*, https://doi.org/10.1007/978-3-031-65262-2_47

Bibliography

BAKUNIN, MIKHAIL, Letter of 16 December 1869 to Nikolaj Ogarev, in COLELLA, CARMINE, Marussia Bakunin, una rilettura aggiornata della vita e della carriera, Acts of the Accademia pontaniana, vol. LXIII, Naples, 2014

BARTOCCI, CLAUDIO, Dimostrare l'impossibile. La scienza inventa il mondo, Milan, Raffaello Cortina Editore, 2014

BENDA, JULIEN, Il tradimento dei chierici, Turin, Einaudi, 2012

BENZONI, GIULIANA, La vita ribelle, Bologna, Il Mulino, 1985

BILOTTA, GIUSEPPE, Renato Caccioppoli per immagini, Naples, Istituto grafico editoriale italiano, 2005

CAIANIELLO, EDUARDO RENATO, Renato, amico e maestro, Il Mattino, 3 November 1992

CALASSO, ROBERTO, Ciò che si trova solo in Baudelaire, Milan, Adelphi, 2021

CARBONE, LUCIANO-CARDONE, GIUSEPPE-PALLADINO, FRANCO, Una conferenza stenografica di Renato Caccioppoli, in Rendiconto dell'Accademia delle scienze fisiche e matematiche, vol. LXIV, Naples, Liguori, 1997

CARBONE, LUCIANO-TALAMO, MAURIZIO, Caccioppoli intimo, in Rendiconto dell'Accademia delle scienze fisiche e matematiche, vol. LXXVII, Naples, Liguori, 2010

COLELLA, CARMINE-CARBONE, LUCIANO, Assilli e disagi nella discendenza napoletana di Mikhail Bakunin, in Acts of the Accademia pontaniana, vol. LXX, Naples, 2022

COLELLA, CARMINE-DRITSAKOU, MARIA GLYKERIA, Ritratto inedito di Maria Bakunin quale si disvela dall'esame della lunga corrispondenza con Max Nettlau, in Acts of the Accademia pontaniana, vol. LXIX, Naples, 2020

© The Author(s), under exclusive license to Springer Nature Switzerland AG 2024
L. Foschini, *The Friction of Life*, https://doi.org/10.1007/978-3-031-65262-2

COMBATTENTE, ETTORE, Rosso antico. Memorie di vita di sezione e di sinda-cato, Rome, LiberEtà, 2009

CROCE, ELENA, La patria napoletana, Milan, Mondadori, 1974

DE CRESCENZO, LUCIANO, Storia della filosofia greca, Milan, Mondadori, 1987

ESPOSITO, SALVATORE, La cattedra vacante. Ettore Majorana: ingegno e misteri, Naples, Liguori, 2021

FERGOLA, PAOLO (edited by), Sulla figura di Renato Caccioppoli, Università degli studi di Napoli Federico II, Dipartimento di matematica e applicazioni Renato Caccioppoli, RISMA, 1994

GIDE, ANDRÉ, Diario, vol. II, 1926-1950, Milan, Bompiani, 2016 Rencontre à Sorrente, Paris-Alger, L'Arche, 1944

GRAMICCIA, ROBERTO, La regola del disordine, Rome, Editori Riuniti, 2004

GRASSI, GINO, Cuore anarchico e cervello di matematico, Il Mattino, 8 July 1982

GRASSO, LUCIANO-VAGLIO, PAOLO (edited by), Tutta Napoli, Naples, Deperro, 1959

GUARINI, RUGGERO, Il matematico dandy che fece i conti senza il comunismo, Il Giornale, 15 February 2004

GUERRAGGIO, ANGELO-NASTASI, PIETRO, Renato Caccioppoli a 100 anni dalla nascita, Berlin, Springer Verlag, 2004

GUIZZI, FRANCESCO, Come stelle fisse, Naples, Tullio Pironti Editore, 2010

HERLING, GUSTAW, Diario scritto di notte, Milan, Feltrinelli, 1992

LA CAPRIA, RAFFAELE, Ferito a morte, Milan, Mondadori, 1984

LA VIA, PIETRO, Caccioppoli e Gide a Sorrento, Il Mattino, 15 July 1979

LEVESQUE, ROBERT, Journal inédit, in Bulletin des amis d'André Gide, vol. XXI, 100, Paris, October 1993

LOMBARDO RADICE, LUCIO, Ricordo di Caccioppoli, l'Unità, 12 May 1959

MARCENARO, GIUSEPPE, Il ritorno di Giovanni Ansaldo, in Nuova Storia Contemporanea, year XI, n. 6, November-December 2007

MARINO, LUIGI (edited by), Napoli-Mosca. L'Italia-Urss di Napoli nei duri anni della guerra fredda 1946-1961, Naples, De Frede Editore, 2019

MAROTTA, GERARDO, Solitudine del genio, Il Mattino, 8 July 1982

MARTONE, MARIO-RAMONDINO, FABRIZIA, Morte di un matematico napoletano, Rome, Ubulibri, 1992

MELVILLE, HERMAN, Naples in the Time of Bomba, edited by Gordon M. Poole, Naples, Alessandro Polidoro Editore, 2019

MONGILLO, PASQUALINA, Marussia Bakunin. Una donna nella storia della chi-mica, Soveria Mannelli, Rubbettino Editore, 2008

ORTESE, ANNA MARIA, Il mare non bagna Napoli, Florence, Vallecchi, 1967

PACE, GIOVANNI MARIA, La morte è una formula matematica, L'Espresso, 20 May 1979

PALERMO, MARIO, Memorie di un comunista napoletano, Parma, Guanda, 1975

PARLANGELI, ANDREA, Uno spirito puro. Ennio De Giorgi. Genio della matematica, Lecce, Edizioni Milella, 2015

PEROZZI, ETTORE. Il cielo che ci cade sulla testa, Bologna, Il Mulino, 2016

PEROZZI, ETTORE-CELLETTI, ALESSANDRA, Asteroid (9934) Caccioppoli: What's in a Name?, in Modern Celestial Mechanics: from Theory to Applications, Berlin, Springer Science & Business Media, 2002

POMILIO, MARIO. In ricordo di Luigi Incoronato. Quattordici anni dopo, Naples, Guida Editori, 1981

PUNTILLO, ELEONORA, Capri. Storia, case e personaggi attraverso la vita e l'opera di Carlo Talamona 1903-1975, Naples, Grimaldi & C., 2013

RAMONDINO, FABRIZIA, Althénopis, Turin, Einaudi, 1981

RAMONDINO, FABRIZIA-MüLLER, ANDREAS FRIEDRICH, Dadapolis. Caleidoscopio napoletano, Turin, Einaudi, 1995

REA, ERMANNO, Mistero napoletano, Turin, Einaudi, 2003 Renato Caccioppoli. La Napoli del suo tempo e la matematica del XX secolo, Naples, La Città del Sole, 1999

RIONERO, SALVATORE, Alcuni aspetti della Scuola matematica napoletana: fantasia matematica e proiezione internazionale, in Rendiconto dell'Accademia delle scienze fisiche e matematiche, vol. LXIII, Naples, Liguori, 1996

SBORDONE, CARLO, Renato Caccioppoli nel centenario della nascita, in La matematica nella società e nella cultura – Bollettino dell'Unione matematica italiana, Bologna, Zanichelli, August 2004 Il pensiero matematico del XX secolo e l'opera di Renato Caccioppoli, Naples, Istituto italiano degli studi filosofici, 1989

SCHIFANO, JEAN-NOËL, Le coq de Renato Caccioppoli, Paris, Gallimard, 2018

SCIASCIA, LEONARDO, La scomparsa di Majorana, Milan, Adelphi, 1997

SCORZA DRAGONI, GIUSEPPE, Renato Caccioppoli, Rome, Accademia dei Lincei, 1963

Si è ucciso ieri il prof. Caccioppoli, il Roma, 9 May 1959

TOMA, PIERO ANTONIO, Renato Caccioppoli. L'enigma, Naples, E.S.I., 1992 Uno scienziato napoletano, Renato Caccioppoli, directed by Marussa Gravagnuolo, texts by Antonio Ghirelli. Documentary, Rai 3, 1982

VALERIO, CHIARA, Lo scienziato Erik Asphaug, Galilei e la luna, la Repubblica, 16 June 2021

VALLETTA, CORRADO, Renato Caccioppoli. Nei ricordi del Maestro Roberto De Simone, Periodico di matematica per l'insegnamento secondario, edited by Ferdinando Casolaro, Franco Eugeni, Luca Nicotra, Roseto degli Abruzzi, AFSU (Academy of the philosophy of human sciences), June 2021

VESENTINI, EDOARDO, Renato Caccioppoli e l'analisi complessa, in Ricerche di matematica, suppl., vol. XL (1991), Acts of the Conference in Naples (1989) The letter from Paola Masino to her mother is held in the Fondo Masino, Archivio del Novecento, Università La Sapienza, Roma; in MASINO, PAOLA, Io, Massimo e gli altri. Autobiografia di una figlia del secolo, Milan, Rusconi, 1995. The letter from Sofia Caccioppoli Bakunin to Adele Croce is held in the Fondazione Biblioteca Croce, Archivio di Benedetto Croce, carteggio per anno e corrispondente, anno 1943, n. 207

Index